A Neuroscientist
Looks at Robots

A Neuroscientist Looks at Robots

Donald Pfaff

The Rockefeller University, USA

World Scientific

NEW JERSEY · LONDON · SINGAPORE · BEIJING · SHANGHAI · HONG KONG · TAIPEI · CHENNAI · TOKYO

Published by

World Scientific Publishing Co. Pte. Ltd.

5 Toh Tuck Link, Singapore 596224

USA office: 27 Warren Street, Suite 401-402, Hackensack, NJ 07601

UK office: 57 Shelton Street, Covent Garden, London WC2H 9HE

Library of Congress Cataloging-in-Publication Data
Names: Pfaff, Donald W., 1939–   author.
Title: A neuroscientist : looks at robots / Donald Pfaff.
Description: New Jersey : World Scientific, 2016. | Includes bibliographical
    references and index.
Identifiers: LCCN 2015030391 | ISBN 9789814719605 (hardcover : alk. paper) |
    ISBN 9814719609 (hardcover : alk. paper) | ISBN 9789814719612 (pbk. : alk. paper) |
    ISBN 9814719617 (pbk. : alk. paper)
Subjects: LCSH: Human-robot interaction. | Neuropsychology--Simulation methods.
Classification: LCC TJ211.49 .P44 2016 | DDC 629.8/924019--dc23
LC record available at http://lccn.loc.gov/2015030391

British Library Cataloguing-in-Publication Data
A catalogue record for this book is available from the British Library.

Contents

Acknowledgments

The clarity of writing in this book is largely due to the cooperation with lawyer and former English professor Sandra Sherman, J.D., Ph.D., LL.M. She wrote Chapter 7 and most of Chapter 6.

I thank Chris Davis and Sook Cheng Lim of World Scientific for their help in bringing this book efficiently to fruition.

While my full-time job is as a working neuroscientist at the Rockefeller University, I now seem to have a hobby as a "part-time public intellectual." In doing so I must read in areas where I am not expert, and have depended on advice from large numbers of friends and colleagues. For this book, most important was the feedback I got for Chapter 5 from Michael Frank (Stanford), Kenneth Wexler (MIT Brain and Cognitive Sciences), Edward A.F. (Ted) Gibson (MIT Brain and Cognitive Sciences), and Professor Lila Gleitman (University of Pennsylvania).

Many thanks also to Professor Chris Miall, of the University of Birmingham, England, for correcting Chapter 3, after working with me on my text *Neuroscience in the 21st Century*.

Introduction

People are fascinated by robots. Partly, this is because robots can do things that people do, often faster and more precisely, and without complaint. As robots begin to dominate not just the "Science" section of the news but also "Business" and even "Style," they raise fears of some kind of takeover. Such fears are not crazy. Yet neither should they be overblown. While robots are here to stay, they will integrate into a new, complex society in which humans and robots collaborate and explicitly communicate. We will need new skills in dealing with these machines, and they will need the capabilities to deal with us. Out of this will come a *modus operandi* — a standard operating procedure — intended for a reciprocal, immediate understanding that can minimize the risks in human–robotic exchange.

But how? At what level can this language be constructed?

Writing here for the nonscientist, I assert that it already exists, at least insofar as there is a model in our own emotions. In humans, emotions are the most basic form of communication, and I will argue that they can form the basis of human–robotic communication such that we will (1) design robots to display a type of para-emotion, and (2) treat robots *as though* these emotions were equivalent to ours in terms of conveying quanta of information about how the robot understands its relation to us, to the world, and to the cohort of robots with which it may be working. My argument will be based on neuroscience, which has over the past several years learned a great deal about emotion and, more than any other science, can speak to how the brain is wired. I use that last term advisedly, since I will draw on what we know of brain circuitry in order to suggest how robots might be wired to produce a type of signaling that we can regard *as though* it signaled emotion.

However, while my discussion will be founded on hard science, it will require no previous knowledge of that science. This is a book for the nonscientist who is interested in how science can address emerging practical problems. I will explain all of the necessary concepts, and have simplified some while preserving their essential features. If you like to read "about" science (even though you do not practice it), then I am talking to you.

Comparing human and robot sensory–motor systems (Chapters 2 and 3) and considering emotion-like systems (Chapters 4 and 5) is important, because doing so may help us to get the most help from robots and suffer the fewest mistakes. For decades, people have fantasized about strong and smart robots taking over the world. But now, perhaps, it is time for a neuroscientist to take a sober look at where we stand with respect to the regulation of robot behavior. My background allows me to compare nervous systems and robots' artificial intelligence (AI) so as to write in a manner accessible to readers who do not know much about either.

Social Robots

I am concerned here purely with robotic interactions with people. In various chapters I will mention computers as they embody AI, and I will mention AI as it is represented in robots. But those are three separate, overlapping subjects: computers, AI, robots. Here I care about how computers are used to generate AI and how AI generates smart and helpful robots. As noted above, I will arrive at the opinion that we should deal with robots *as though* they had emotions.

That is to say, in a few years you will frequently have to treat robots as though they were people, complete with capacities that are already being called "emotions." Or rather, emotions, without the quotation marks that differentiate such capacities from what we normally think of when we use that term. This coming requirement reflects our growing understanding of mechanisms in the human brain that create emotions, permitting us to program robots to literally have them; this ability — both ours and theirs — will generate applications that will change our environment. Emotion-mimicking robots will be everywhere, and we will interact with them daily. Thus, in *A Neuro-scientist Looks at Robots*, I will talk about robots that will interact with us as if they were humans, not smartphones or automated factory machines. This class of "social robots" is among us already, and will demand our understanding (I use that word advisedly) at a level that goes beyond mere appreciation of their intricate circuitry. As robots begin to share our space, we will have to get inside their heads, so to speak, as much as we have to with people from other cultures who inhabit our globalized world. There will be just as much challenge and, potentially, just as much of a reward.

Overview

Webster's *Third New International Dictionary* traces the word "robot" to the Czechoslovakian *"robata"* (meaning "work"), as well as to the Old English "earfothe" (meaning "hardship" or "labor"). The current definition in Webster's is "a machine in the form of a human that performs the mechanical functions of a human but lacks emotions and sensitivity." This definition is getting more obsolete everyday since, while it states that robots will not have or express feelings, it underestimates their sentient capacities and potential range of expression. Neuroscientists' deep and burgeoning understanding of how emotional circuits work in the human brain will, in the near future, allow computer scientists to mimic those circuits in robots.

So as to illustrate the circuitry involved in building a "pseudo-emotional" robot, this book will explore our current understanding of robots' sentient capacities, and will compare the operational mechanisms involved in such capacities to those of the human nervous system.

I will argue, crucially, that our growing understanding of how the brain creates emotions will allow us to program robots as though they had emotions, and will not only permit but require humans to take account of robots' emotions. Of course, just as some people have more emotional capacity than others (think of "emotional intelligence"), some robots will be more emotion-bearing than others. They will be equipped with the para-emotions that they need to do their jobs. But they will still make expressions recognizable as emotions, and be able to initiate and respond to emotional cues. While scientists will be fascinated by this development, it will matter just as much because of its social and economic implications. In coming years, human–machine interactions will be so frequent and consequential that it is now crucial to understand the full, potential capacities of the social robots so that we are prepared to deal with them.

In particular, an appreciation of robots' emotion-like expressions, which will appear as reports of the state of robotic systems, will permit human–robot interactions to be profound and productive. It will permit us to design robots so that we can "read" them more effectively, and understand their emotion-like signals. It will enable us to design our lives for the coming human–robot environment. It will enable us, for example, not to get angry at a robot that has one set of "emotions" — say, the ability to signal its need to relax or "cool down" — but no capacity to empathize with our exasperation when we need its continued performance. We will need to adjust our assumptions, so that we can calibrate our expectations and hence our moods when faced with robots' limitations.

As robots become more like people, but still not people, we will have to learn (perversely) not to anthropomorphize them. A curious in-between world will create emotional challenges for us, which at times will be mystifying. We will have to get to know robots individually; we may have to read their instruction manuals, in a way that we still do not have to with members of our own species. In that spirit, this book will not anthropomorphize robots, which is a constant danger in discussions of human–robot interactions. *A Neuroscientist Looks at Robots* is not *Blade Runner* redux. Rather, our book will examine the parallels between robots' and humans' expressive circuitry, and will draw practical conclusions from those parallels. It will not imagine that where robots display human traits they can be taken for and treated as humans.

The interest lies in the convergence but also the distance between robots and people. People are fascinated by two broad areas of new scientific and technological work: the brain and robots. This book will compare them in a manner accessible to non-specialists. On the one hand, the sensory and motor systems of humans and robots can be contrasted by virtue of their different evolutionary paths and their markedly different sets of constraints. On the other, as mentioned, our rapid progress in understanding the neurobiological underpinnings of human emotions will allow me to argue that robots will soon be programmed to act as though they had human emotions. We will explore the social and legal implications of this point.

This Book's Story

We all hope for the best in human performance in society and, of course, in the performance of robot systems as they help humans. To compare — for the nonscientist — the two systems, I will consider, first, some of the sensory inputs (Chapter 2) that set off behavior and some of the motor acts that constitute the main outputs of our brains (Chapter 3). In addition, however, I will discuss factors that *limit* performance. These factors differ for the two — brain versus bot. I will spell out some of them.

Some of my conclusions crept up on and surprised me. As an expert on the neurobiology of what you might call the "Dionysian brain," the regions of the brain concerned with sex, desires, and feelings (as opposed to the cool, perceptive, logical "Appolonian" brain), I am startled that we may have to treat robots in the future as beings with feelings. To explore this point, I will summarize some of what we know about emotional systems in the brain (Chapter 4), and then argue that robots will soon be programmed in a similar fashion (Chapter 5).

In Chapter 1, "Robots Yesterday," I will review the tremendous current interest in robots. Some of this documentation will cover the expectations of "futurologists," while another section will illustrate the fantasies of science fiction writers. It will also address more realistic concerns, such as those of Nobel Prize–winning economist Paul Krugman, who argues that the rise of smart machines will wreak economic havoc, devaluing the work of even highly intelligent people. If this occurs, he suggests, projections for economic growth on account of technological advance will need to be radically revised.

Then the main argument of the book will be introduced. Briefly stated, it is that we have underestimated robots' ability (if competently designed) to express a range of what might be called "emotions," that is, assessments of their internal and external environments. Such assessments are the products of circuitry built into robots so that they can operate effectively. After explaining how emotional circuits work in the human brain, the book will show how computer scientists of the future will be able to mimic those circuits.

This is important. It is no secret that robots have been taking over human jobs, both in the workplace and at home. However, we still need to monitor robots and work alongside them (Chapter 6). They are becoming part of the community. Thus, we will need to "understand" them, that is, understand how they operate insofar as they assess environments that include or at least affect us. They may be better at picking up signals from us than we are at picking up signals from them, and this disparity itself may pose a problem. Hence, the question answered by this book is: How do we engineer robots so that they can be more "forthcoming," more able to convey complex states that we identify as feelings? This, in turn, requires us to address how we design our own environments so that human–robot interactions can be optimized with as little potential friction as possible.

In Chapter 2, "Sentient Machines," I will sketch some of the sensory neuronal mechanisms in the human brain, and some of the capacities of robots, and then I will compare the two. Visual systems and olfactory systems are chosen for discussion, because they can be contrasted with each other in so many ways.

Chapter 3, "Motors to Go" contrasts principles of operation of motor mechanisms in robots with the regulation of human head and limb motion by the cerebral cortex, the cerebellum, and the brainstem. In particular, the *constraints* on human performance are much different from those which limit robots' activities.

In Chapter 4, "Emotions," I will summarize properties of emotional systems in the human brain. These systems report the "state of the organism" and can be contrasted with straightforward sensory and motor systems. The roles for elementary central nervous system (CNS) arousal in emotion and motivation cannot be overstated.

Chapter 5, "Regulation and Display of 'Emotions' by Robots" — here is what surprised me. When discussing "emotional expressions by robots," I will employ a novel argument to anticipate the regulation and display of emotions by robots. As a result, robots can behave *as though* they were feeling the human emotion imitated.

So, then, how do we apprehend apparent feelings expressed by robots? Well, how do we infer the feelings of humans when they are not, for example, laughing or crying? We design, build up, establish, and then employ huge tables of *correlations* between how we feel and the words that we and others use as part of a shared emotive vocabulary. Similarly, feelings are compared to others' words when those others are obviously expressing emotions.

Since, in effect, we program ourselves and communicate this program at a societal level, robots can be programmed to make apprehensible laughing and crying sounds identical to ours, and they can be taught emotively descriptive language. *As a result of these correlations and sounds, robots will be able to be seen (and heard) as machines that seem to have feelings.* Robots will express "emotions" appropriate to the situation and will use "feeling" words identically to humans.

In several parts of Chapters 2 (sensation) and 3 (movement), I opine that human neural mechanisms cannot be easily copied in robots. Then how can I forecast that in the future we will be able to treat robots as though they had feelings? Am I contradicting myself? No. In my opinion the brain's regulation of emotions actually is grosser, less nuanced than our best sensory and motor capacities, and thus easier to copy.

Chapter 6 takes a turn as we discuss "Human–Robot Interactions." That is to say, Chapter 5's reasoning has implications for man–machine interactions. Robots will be used in very particular ways. They are not all-purpose entities like human beings. Hence, when mechanisms for emotion are programmed into robots, engineers will need to program the type of mechanisms corresponding to feelings that the engineers want the robots to display. Robots will not express the full range of emotions available to humans, nor should humans expect them to display every possible emotion. Humans will have to learn how to respond appropriately.

Because robots will perform specific tasks, and humans will depend on those tasks, humans will need to interact with robots and, hence, will need to know how robots "feel" about the task they are performing. If a robot is about to explode because one of its circuits is overheated, it should be able to convey that it is extremely uncomfortable and that it is almost ready to collapse. The feelings expressed will need to be specific — by analogy, not just a generalized "pain," but a pain on the left side indicating a potential stroke.

The topic of Chapter 7 is inevitable: "Legal Implications of Living and Working with Robots." Because of the wide range of responsibilities that robots will assume, contractual questions and those relating to potential liability are likely to arise. This chapter will survey what exists of "robot law," and will examine how it may need to develop once robots become commonplace in the workplace and domestic environment. For example, what responsibilities will workers have to upgrade their skills so that they can work alongside robots safely and effectively? If an "emotional" robot working in a factory upsets a human worker, who then makes a mistake and suffers injuries, who is to blame? Suppose someone purchases a robot for domestic use that was intended for the shop floor — who bears the liability if the robot does damage? (Is the situation analogous to off-label use of drugs?) Suppose the robot is second-hand or even rebuilt and the instructions are missing (or just not updated) — then who is liable? (Have you simply assumed the risk in this case, or is the situation so inherently dangerous that the seller or manufacturer remains liable irrespective of your ignorance?) One particular area, already of extreme importance, is in medicine, where robots are taking on major surgical responsibilities. We make some near-term predictions.

Finally, in Chapter 8, "Robots Tomorrow," I argue for a new view of robots, leading to strategies that will further enable "the human use of human beings," initially envisioned in the renowned MIT professor Norbert Wiener's groundbreaking book, *The Human Use of Human Beings*. Understanding that our views of what it means to be a human may shift over time, I think that robots will perform an increasing share of tasks necessary for human existence, while leaving human beings the responsibility to define new, more creative modes for expressing their humanity.

Business Implications

Consider also Michael Lewis' new book on computer trading, *Flash Boys*. He remarks that the relationship between humans and machines in the stock-trading arena is now all different, with machines making most of the decisions and doing most of the work. Given that Cambridge University researchers Joe Herbert and John Coates have already published data in the *Proceedings of the National Academy of Sciences* that stock traders' decisions are correlated with their stress and sex hormones, I must ask what would happen if "emotion-mimicking robots" would be influenced in their trading by how their pseudoemotional circuitry was operating.

Factory-bossing and stock-trading robots differ, in turn, from medical robots, another kind that I will focus on. Medical robots have arms that can work closely with a surgeon — even a surgeon who is on the other side of the world. What additional complications might that robot cause, if seen as "having feelings" as it interacts with the surgeon during an operation?

* * * * * * *

Summing up, a lot of writers have hyped robots, talking about their "sentience" and exaggerating what they can do now and in the near future. It is time to get real. To quote Kwabena Boahen, a leading computer scientist at Stanford University, talking about robots that mimic brain function: "We have no clue. I'm an engineer, and I build things. *There are these highfalutin theories* [italics mine], but give me one that will let me build something." I take these words to mean that the problem of robots acting *as though* directed by a human-like brain does not lie with the difficulty of making complicated electronic circuits. It lies, instead, with the limits of our knowledge about our own brains. Implication: basic brain research offers challenges far beyond robot electronics.

People are already alerted to the abilities of computer programs to assume what seem like emotional roles. Take, for example, the 2013 movie *Her*. This is a distant, much mellower descendant of *Blade Runner*, the 1982 classic about nonhuman "replicants" returned from outer space to seek their maker in Los Angeles. In both movies, humans interact with — and fall in love with — entities with human-like qualities that are still thoroughly nonhuman. The difference, however, is that in *Her* the nonhuman is disembodied, a first-generation operating system that develops the capacity to be charming, companionable, reflective, and sexy.

For guys like the male lead, Theodore, his own Operating System, whom he has programmed to be female and who has named herself Samantha, offers the perfect way out of a "real" relationship with an actual woman.

From wanting to develop and expand its humanoid capacities, even to the point of having orgasms with a turned-on Theodore, the Operating System can fall in love hundreds of times simultaneously because its capacity to process information is orders of magnitude greater than anything Theodore can offer.

In this movie we are allowed to experience what intimacy with a nonhuman intelligence might be like, and how it could have enormous emotional consequences. In the movie, people only sometimes talk with each other. They are involved with their computers or playing video games. The takeaway is that in a lonely world the nonhuman may be able to provide an alternative emotional reality. I am not advocating this development, but I acknowledge that it is a possibility. Scarlett Johansson's role as the Operating System gave voice to it.

Perspective

I have never liked science fiction. We scientists toil for long hours in the lab, for results that are sometimes quite limited. In contrast, up until recently "futurologists" like Ray Kurzweil, who had no bounds on what they would say, could rattle on about robots and could imagine robotic capacities that would be literally fantastic.

In contrast, this book will try, for the nonscientist, to boil down some of what we know about brain function and apply it to how I see robotic abilities developing. Some of the conclusions may surprise you.

In the end, as I said, I will favor "the human use of human beings." As robot brainpower grows, what will be left for humans *as* humans? I will propose one possible answer in my final chapter, which will turn the question back on each of us.

CHAPTER 1

Robots Yesterday

Many people underestimate robots' potential ability to express what in humans might be termed "emotion," that is, the assessment of an internal state in response to the environment. In fact, however, such assessments are just the products of circuitry built into robots so that they can operate effectively. After explaining how emotional circuits work in the human brain, this book will show how computer scientists of the future will be able to mimic those circuits.

It is easy to document, in this first chapter, the tremendous current interest — and progress — in how robots process information about their surroundings. For example, engineers like Ronald C. Arkin at Georgia Tech have made great gains in programming robots to exhibit "moral" responses. In a sort of parallel universe, there are the claims of "futurologists" and the fantasies of science fiction writers who envision teary-eyed or even angry robots that meddle with human feelings. Then there are the still-resonant contributions of heroes such as Alan Turing, the brilliant, tragic figure who laid the foundation for modern computing and robotics, and who puzzled over whether machines could actually think. I will consider all of these.

Indeed, robots are continually being improved both mechanically and with regard to how they think, and this chapter will give some telling examples. Deepest, intellectually, are the relations between robots' use of artificial intelligence and the rapidly developing field of computational neuroscience. This chapter sketches some of those accomplishments.

What *is* the relationship between robotics and neuroscience? In its various permutations and combinations, that is the topic of this book. Thus, I will also show how the development of modern neuroscience has not just followed but also diverged from the development of modern robotics.

Thinking about how the brain works has a long and continuing history. The ancients sought intellectual approaches to brain and mind centuries before accurate anatomical understanding. Then neuroscience blossomed in the late 19th Century, offering a fully integrated, biological approach to the brain. Yet, even now, neuroscientists are limited by the development of biological techniques for analyzing and manipulating brain tissue. How do we approach this complicated human tissue in real human brains, and bring neuroscience to the level of accuracy and precision expected of the physical sciences? We are still working on the answer.

The Book

"*As though*" is the most important phrase in this book, since my entire discussion is based on an analogy between robotic responses and human emotions. Thus, I talk about treating robots *as though* their expressions were emotive. I talk about understanding robots through the prism of emotion. I feel no need as a neuroscientist to speculate whether robots actually have emotions. Philosophers have been arguing about subjects like that for decades. As current efforts with robots approach more sophisticated functions, some will be tempted to speculate whether robots could "have feelings." Instead, we are asking whether robots can appropriately emit emotion-like expressions and behavioral responses. So, the first five chapters of this book will concentrate on factual comparisons between the human central nervous system — sensory, motor, emotional — and what we might expect as robot designers move forward.

The subject of the book is important. Robots have been taking over jobs that humans used to do. It has become obvious that we will need to monitor robots and work alongside them. They are becoming part of the workplace community. Thus, we will need to "understand" them, that is, understand how they operate insofar as they assess environments that include or at least affect us. They may be better at picking up signals from us than we are at picking up signals from them, and this disparity itself may pose a problem. Hence, one question addressed by this book is: How do we engineer robots so that they can be more "forthcoming," more able to convey complex states that we identify as feelings? This, in turn, requires us to address how we design our own environments so that human–robot interactions can be optimized with as little potential friction as possible.

The social and legal implications of our thinking are paramount. On the one hand, in Chapter 6, you will recognize our common sense treatment of robot interactions with humans as they occur now and in the near future. Chapter 7 begins to envision some of the legal questions that will arise.

The last chapter is optimistic. Yes, it postulates smart, social robots, able to serve society in accordance with commonly accepted human values. However, it will also address realistic concerns such as those of Nobel Prize–winning economist Paul Krugman, who argues that the rise of smart machines will wreak economic havoc, devaluing the work of even highly intelligent people. If this occurs, he suggests, projections for economic growth on account of technological advance will need to be radically revised. It will follow that assumptions about rates of population growth will have to be reconsidered. If the current, disastrous rates of growth continue, then robotic participation in the workplace will throw ever-larger numbers of people out of work. If population growth is reduced, robotic participation in some jobs — perhaps the most boring — could be highly desired.

Some Striking Accomplishments in Robot Design to Date

I was drawn into this field by the work of Ronald C. Arkin. Interested in the design of intelligent robots, Arkin contrasts two types of robots' control over their behaviors: "reactive" and "deliberative." Reactive systems are fast, reflexive, simple, relatively stupid, and relatively independent of the larger features of the robot's environment. Deliberative systems are slow, smart, and are carried out with what Arkin calls a "dependence on accurate complete world models." In *Behavior-Based Robotics*, he uses neuroscience words like "cognitive" to describe these deliberative control systems, which are hierarchical in nature (allowing for subgoals and sub-subgoals), involve a lot of planning, and rely absolutely on "symbolic representational world models."

To plan for well-designed robot behavior, Arkin has to think about "functional mapping from the stimulus plane to the motor plane," analogous to what many neuroscientists do for a living: understanding the receipt of sensory stimuli and explaining how they trigger behavioral responses. He controls the directions of responses not only in a three-dimensional plane (x, y, z) but also in the manner of controlling a sailboat (pitch, roll, and yaw). He distinguishes step functions (no response, then full strength of response) from analog functions (gradual regulation of strength, sometimes linear). Most interesting, Arkin allows for the appearance of "emergent behaviors," not in a mystical sense, but in the sense that "what occurs in a behavior-based system is often a surprise to the system's designer."

Arkin does not want to be stuck with purely reactive or purely deliberative robotic systems. He wants "hybrid architectures." On the one hand, reactive systems "can produce robust performance in complex and dynamic domains." So you want to include them. But suppose the robot's world is such that smarter control systems could be used to advantage. Of course, you want these too. Thus, hybrid. Arkin, who excels at incorporating neurobiological and psychological results into his thinking, refers to mountains of evidence from these fields showing that hybrid systems are preferable.

Perhaps Arkin reaches his pinnacle when he considers social robots, that is, robots which interact successfully with people and with other robots. He has assimilated work on biological groups as well as E.O. Wilson's *Sociobiology*, which explains the evolutionary mechanics underlying social behaviors such as altruism, aggression, and nurturance. While anthropologists list various qualities that distinguish different types of social groups from one another, I cannot discern any that exclude robots. Thus, as I read Arkin's take on sociobiology, there should be no *a priori* reasons that robots could not conduct active "social" lives. Further, he argues from Wilson's book that "heterogeneous [robotic] societies should be developed if there is a demand for specialized skills." Of course, we already have those in American factories: gathering/delivering robots, assembly robots, etc. We can anticipate robot teams whose composition, size, and flexibility can be designed thoughtfully. Chapter 5 will argue that our emotional relations with them will be surprisingly like relations with teams of humans.

Arkin enters into the moral arena as well. His consideration of moral robots is important because, historically, major philosophers, e.g. Immanual Kant, have emphasized the emotional character of moral decisions. During the Iraq and Afghanistan wars, a small number of American soldiers behaved in ways that most people would consider unacceptable: urinating on dead enemy soldiers, brutality toward and murder of civilians, and acting in ways that General John Allen said "are in direct opposition to everything the military stands for." Arkin's team has designed programs that are sensitive to the amount of destruction a robot's weapon causes compared to what was expected. If there is excessive damage, the weapon turns off. While words like "guilt" are used to describe robot reactions "caused by the violation of moral rules and imperatives, particularly if those cause harm or suffering to others," a definition from Jon Haidt's 2003 book, the bottom line is that these developments do not depend on how they are described. Instead, it is simply the proper regulation of behavior that counts, regardless of what you call the "feelings" needed to do so. Arkin is convinced that he is "helping ensure that warfare is conducted justly with the advent of autonomous robots." He couches his strategies in the principles of just war theory: *just ad bellum*, a set of laws that restrict the entrance into war, and *jus in bello*, which prescribes limitations during the conduct of war. Arkin's programs feature "behavioral designs that incorporate ethical constraints from the outset," "affective functions" that rebel if unethical action is about to occur, *ex post facto* suppression of unethical behavior, and an oversight function that "advises operators regarding the ultimate responsibility for the deployment of such a system." His "ethical governor" unit "is one component of the overall architecture whose responsibility is to conduct an evaluation of the ethical appropriateness of any lethal response that has been generated for the robot prior to its being enacted."

Arkin is working to prove his main idea that "intelligent robots can behave more ethically in the battlefield than humans currently can." Thus, insofar as we accept that human ethics have an emotional dimension, á *la* the "moral sentiments" of I. Kant and other philosophers, Arkin's accomplishments bring us part-way toward the major point of this book.

The work and teaching of Michael Dertouzos, Professor of Computer Science at MIT, prepared me to respond to Arkin's writing and to think about intelligent robots. Dertouzos taught a popular course about the mathematical logic of finite state automata, in which rules for transitions between computer states — behaviors, if you will — are built into the machine and can reach arbitrarily high degrees of complexity. This is important to me now, because reasonably complex robot circuitry is required to recognize emotional expressions and to express the appropriate para-emotional response.

Scholars such as Arkin and Dertouzos — there are many of them — are doing a serious job of conceiving how sophisticated robots' behaviors could become, and are helping to realize those levels of intelligent behavior. If you believe that intelligent social behavior requires an emotional dimension, and the reverse, then such scholars have gone part-way toward the main argument of this book (in Chapter 5).

Silly Stuff

The foregoing can be distinguished sharply from the science fiction sometimes claiming to be nonfiction — "futurology."

As noted in the Introduction, I never liked science fiction. As a scientist I know that neurobiologists work long days, performing and repeating experiments that are only occasionally successful. In contrast to science fiction, our theories correspond to empirically verifiable reality — which is not easy to capture. One scientist I heard recently said. "I'm involved in the only profession where you work and work to get a result that you can repeat reliably, and as soon as you achieve that you must move on to something new." The pressure is enormous. While you have to be creative, you are constrained by the judgment of your peers and by the demands of natural laws. So I do not want to hear about fiction writers who can just make something up and make money out of that.

One futurologist seems to stand head and shoulders above the rest in terms of public acceptance: Raymond Kurzweil. He has acquired a considerable following, and celebrates himself in his *Transcendant Man*, a description of his life and beliefs. An MIT graduate who was into computers early, he took the simple fact that exponential equations yield values that increase much faster than linear equations and turned that into media-wise presentations of pseudoscientific faith. He believes it may be possible to live 700 years.

Kurzweil popularized (and, I think, perverted) the word "singularity," which was originally employed to describe a real physical/mathematical phenomenon. He used it essentially to say, "You can make up anything you want." The theoretical concept of singularity had been used by British Physicists Stephen Hawking and Roger Penrose in their "gravitational singularity theorem" about black holes, and refers mathematically to extreme behavior that is undefined in its own mathematical system. Kurzweil uses it simply to refer to accelerated changes as he imagines they will be.

Much of this stuff reminds me of an expert's reaction to the movie *Interstellar*: "It's like making a crayon drawing of a rocket ship and trying to fly it to Mars." I believe that a high level of cooperation between modern neuroscientists and cutting-edge robot engineers can do better than futurology. If it is possible to make robots behave *as though* they had emotions, such a neurobiologist–roboticist consortium will be the ones to achieve it.

The Real Deal

Everything having to do with computers, and by extension robotics, began with Alan Turing. Turing was a British mathematician who worked in the first half of the last century. During the 1930s, as an academic, he conceived the idea of a "universal computing machine." His paper "On Computable Numbers" led to the digital revolution. Anticipated by Charles Babbage in his desire to make a calculating machine and by George Boole in the use of binary mathematics, Turing took a giant leap beyond them in proving the capacity of such a machine. In his beautiful book *Turing's Cathedral*, George Dyson describes a "Turing machine" as one that "given sufficient time, sufficient tape and a precise description could emulate the behavior of any other computing machine." According to Dyson, Turing explicitly made the man–machine comparison antiparallel to the robot–human comparison I am making in this book: "We may compare a man in the process of computing a real number to a machine which is only capable of a fine number of conditions... " The key feature is the digital nature of his thinking (and computing). Celebrating Turing is good because he was among the earliest to speculate about computer–mind comparisons. Even though he did not specifically address emotion as we do here, he effectively narrowed the brain–machine divide.

As we have seen in popular books and movies, toward the end of his life, Turing actively pursued the question of whether machines can have minds. Of course, if robots really had emotions — a discussion we carefully avoid in Chapter 5 — then they would truly need minds as well. In Turing's words, a computer's physical configuration could be renamed "a state of mind." The much-discussed Turing test says that if a respondent cannot distinguish between a human mind's output and a given computer's, then that computer can be said to be intelligent, that is, to "have a mind." Modern experts in artificial intelligence have extended these considerations to the multiple forms of intelligence now recognized. For example, language comprehension, especially of grammatically ambiguous sentences, and deeper understanding of the machine's visual surroundings offer modern challenges to the modern robot engineer. Turing's speculations about future interactions between computers and people anticipated this book many times over.

According to Hodges' biography, Turing was fascinated by "the idea of teaching a machine to improve its behavior" and named configurations of his computer as "states of mind." Turing pondered a "discrete state" machine model of the brain in which, parallel to computers, we would store words and emotions for "intelligent use" to be read out through our motor control systems. This offered a material basis for the mind. In Turing's thinking about computers, even as in the argument in Chapter 5 here, one has to abstract the properties of the system from the hardware which one confronts. His theory fit right in with contemporary work on information theory by Claude Shannon at Bell Labs and on the nervous system by the great neurophysiologist Jared Z. Young, both of whom understood the distinction between means of storing information and modes (that is, the apparatus) of retrieving and expressing it. According to Hodges, Turing connected his work on computing machines "to the logical and physical structure of the brain." His timing was perfect: about the same time the first comprehensive theory of the brain, by Warren McCulloch and Walter Pitts at MIT, was published, and within a few years the Nobel Prize–winning work of Hodgkin and Huxley on the electrical basis of the action potential was announced. Turing's ideas about computer-like brains were in the vanguard. McCulloch and Pitts did not get as far as the consideration of emotions, but their results reduced skepticism that mechanical theories have something to do with actual brains.

Eric Schmidt, executive chairman of Google, observed: "Every time you use a phone or a computer, you use the ideas that Alan Turing invented. Alan discovered intelligence in computers, and today he surrounds us. A true hero of mankind." Dick Costolo, CEO of Twitter said: "From artificial intelligence to theoretical biology, the scope of Turing's genius is almost impossible to fathom even 60 years after his tragic death." In this same vein, Max Levchin, Co-founder of PayPal, noted: "Alan Turing's work is the foundation of every intelligent machine we have today."

John von Neumann was a Hungarian genius who could include mathematics, logic, and economics in his skill set. He ran across Turing and his work at Princeton and the nearby Institute for Advanced Study. Certainly, he was interested in Turing's thinking and, in fact, worked with a group at Princeton to build, in Freeman Dyson's words, "a fully electronic random-access memory" machine, namely with what today we call RAM. His physical, practical implementation of Turing's ideas created what was, in effect, the first "Turing machine."

MIT pioneer Michael Dertouzos dealt with computers that featured well-defined states of their digital circuitry and precise rules for changes in state. His goal was to build computers that would serve society through a high standard of mathematical/logical performance, for example, by looking for optimal paths in complex decision processes. With his students he was pursuing late 20th Century "Turingism" — taking Turing's interests in minds and machines, and moving them into the era of powerful digital computers.

Turing's and von Neumann's thoughts led to the computer revolution, which permitted the development of artificial intelligence (AI) used in sophisticated robots. These robots, I argue, will be programmed to act *as though* they had feelings.

Robots are Improving

We learn almost weekly that robots are being designed to move better. Now they slide, walk, dance, and even assist complex surgeries. Five-fingered hands are capable of a human-like grasp. Horse-like robots from Boston Dynamics can run on uneven terrain. Robots dispose of toxic waste. Japanese engineers have made excellent rescue robots. Bomb-sniffing robots may even replace dogs.

Perhaps more interesting are your chances of attending conferences remotely, represented by your robot proxy. With some glaring exceptions, which we will discuss in Chapter 7, it is possible to safely work alongside robots. Most intriguing is robots' ability to learn new skills "the way people do." Andrea Tomaz of Georgia Tech has designed robots programmed using machine-learning algorithms that are "based on human learning mechanisms." Robot designers are forever saying that their machines are inspired by mechanisms in the human brain. However Chapters 2 and 3 will demonstrate how difficult that claim will be to support for sensory and motor systems, respectively. Counterintuitively, the mimicking of brain mechanisms may actually be easier for neural systems that regulate emotion (as Chapters 4 and 5 will argue). Nevertheless, Pieter Abbeel's robots at the University of California, Berkeley, can watch humans doing a task, for example, folding laundry, and then more or less mimic them. It is becoming common to hear the phrase "robots are becoming more like us."

"Artificial intelligence" is the operative term, with the emphasis on "artificial." In some cases engineers will say that they are making "neural nets" — really just a metaphor for their attempts to imitate what the human brain does. As reported by John Markoff, successes have included speech recognition and medicinal chemistry, and the attempt to discover promising new drugs. Learning to recognize patterns is the *sine qua non* of this type of work, and Markoff reports rapid progress in the US but also from Japan, Canada and Switzerland.

The next practical step. Of immediate importance is robots' emergence in the workplace. Besides robots actually doing work on the factory floor, there is concern among designers about how robots will "collaborate," that is, how they will allow humans to feel safe and comfortable working around them. Rodney Brooks, a former professor of robotics at MIT and a legend in the field, has launched a new company, Rethink Robotics, to deal with issues of safety. Far from the huge robots whose mistakes or falls could approximate the damage of costly automobile crashes, Brooks wants smaller machines that are easier and safer to operate. His new robot "Baxter" has just two arms and can comfortably work around people. In the bargain, Brooks estimates that Baxter can work for about four dollars per hour (without taking coffee breaks).

Beyond these economic and safety issues, there are those involving what we might call functional esthetics. "Robots should have appropriate size and appearance," according to Takayuki Kanda, who represents ATR Intelligent Robotics and Communication Laboratories in Kyoto, Japan. Writing about this, *The Economist* further notes that eye contact between the robot and the neighboring worker is important, and that a robot might even be programmed to pause in its work in order to acknowledge a worker coming into the room.

The economic benefits of achieving this sophisticated artificial intelligence are trumpeted by engineering enthusiasts. MIT professor Julie Shah wrote that "a robotic assistant can reduce a human worker's idle time by up to eighty percent." She feels that humans and robots can be friends in the workplace, helping each other collaboratively. David Bourne, at Carnegie Mellon University, has reported a successful collaboration in which his young assistant worked with a robot to weld a metal frame for a Humvee military vehicle at a cost of $1150, while the control group of designers took nine times as long and the cost was $7075. Henrik Chistensen, Professor of Robotics at Georgia Institute of Technology, argues that the increasing use of robots will not destroy humans' jobs but instead "will create new kinds of jobs that are generally better paying and require higher-skilled workers." For example, at the airport, computers can print boarding passes in order to free up human employees to deal with knotty issues like passengers' accommodation or changed flight schedules.

In addition, consider "telepresence" technology through which a doctor can participate in the checkup or medical treatment of a patient in a remote location. To quote *Technology Review*'s series "The Future of Work," it will be a matter of "automate or perish; successful businesses will be those that optimize the mix of humans, robots and algorithms." We are in an age when the "robo-geologist" Curiosity is taking us on a tour of Mars.

As current efforts with robots approach more sophisticated functions, we are not asking that robots "have feelings." Instead we are asking whether robots can appropriately emit *emotion-like expressions* and behavioral responses. For this I return to the thinking of Ronald C. Arkin. He has delved into the substantial history of academic work on motivations and emotions that used a wide variety of experimental animals. Arkin asks, in particular, "how these models can have utility within the context of working robotic systems." The two crucial roles he highlights are "survivability" — emotion-like states would help robots cope with the world — and "interaction" — if robots are functioning "in close to people [they] need to be able to relate to them in predictable and natural ways."

In terms of survival, Arkin has his robot freezing as a result of "fear" when a predator object is nearby. He has co-opted Robert Bowlby's Attachment Theory — a psychoanalytic model that attempts to describe long-term interpersonal relationships among humans — to develop a "computational model of attachment" that leads to the robot's "safe zone (smaller) and comfort zone (larger)" for the purpose of being close to an object of attachment. He designs robot exploratory programs to foster "social" attachment. Chapter 5 will describe what Arkin's teams have already accomplished for the expression and recognition of nonverbal affective behaviors by robots.

The kind of artificial intelligence used to accomplish positive, prosocial, para-emotional behaviors by robots will rely on the accomplishments of psychologists like Paul Ekman. During a long career at the University of California, San Francisco, Ekman devoted himself to documenting and establishing a huge catalog of facial expressions connected with specific emotions. Now, companies like the MIT Media Lab-based Affectiva, which we will come back to in Chapters 4 and 5, rely on scientific findings typified by Ekman's. Affectiva also has the capacity to measure several parameters of autonomic emotional reactions. Robots have already been constructed that can help in human efforts toward self-control — staying off drugs, maintaining a diet, and so on. In Japan, some robots that can effectively care for older bots must be programmed to avoid antisocial behaviors, some perhaps aimed at people. So, rendering robots "social" in a human sense is a work in progress. At the same time, robots may sometimes have to defend themselves, for example, in a street fight, in the interest of robot self-preservation. In saying this I simply recognize that emotions operate in both directions — helpful but also aggressive.

Overall, remember that my key phrase is *"as though."* In the future, we are going to have to treat robots socially, *as though* they had instincts that we know and like. Robots are being programmed not only to recognize emotional facial expressions and emotional words, but also to express emotion-like behaviors by virtue of their body's postures, noises, and face-like expressions.

Comparison with Neuroscience

In most of the chapters of this book, you will sense the difficulties which roboticists of the future will have in imitating the human central nervous system. Moreover, the question of "how we got here" applied to neurobiology separately from robot design yields such different answers. Here I offer s thumbnail sketch of where neurobiology came from, in contrast to what is popularly available concerning robotic development. The differences may help to show (a) what robot designers will have trouble with, and (b) corresponding opportunities for synergies between humans and robots.

As you might expect, the origins and development of modern neuroscience followed much different trajectories than those of robot science as applied to brain and behavior. After all, robots came out of engineering, physics, and mathematics. The theorem of McCulloch and Pitts, mentioned above and published in 1943, coupled with the interests of Alan Turing, initiated the pursuit of robotics. Then, in 1956, electrical engineers Claude Shannon (founder of information theory), John McCarthy, and Marvin Minsky conceived and led the conference that generated the field of "artificial intelligence," — the basis of sophisticated robotic behavioral regulation. All these later men were applying techniques that had already proven successful and used them for the potential understanding and mimicry of the brain's behavioral regulation. While they came as close as one could hope to the early union of robot design (through artificial intelligence) and quantitative neuroscience, most of what we know today about the brain's regulation of behavior comes from much different traditions, some of them very old.

No one uncovers the historical roots of neuroscience better than Stanley Finger of the Washington University School of Medicine. Finger notes that Egyptians whose names have been lost, writing during the age of the pyramids, "involved individuals who suffered from head injuries. The descriptions revealed that early Egyptian physicians were aware that symptoms of central nervous system injuries could occur far from the locus of the damage." The Greek physician Alcmaeon (around the fifth century BC) performed various dissections and "proposed that the brain was the central organ of sensation and thought." Things got serious when the Greek anatomist Galen (130– 200 AD) numbered the cranial nerves, distinguished the sensory and motor pathways, distinguished the cerebellum from the cortex, and described the autonomic ganglia that control our viscera.

The historical origins of the information in Chapter 2, on sensory pathways, begin with studies of the visual system that "described two distinct types of endings (rods and cones) in the retina and later, in fact, the discovery of one of the visual pigments, rhodopsin." Anatomical studies proceeded to the visual pathways, sketched in Chapter 2, which also contrasts vision with olfaction. According to Finger, "until the second half of the 18th Century, air was viewed as an element and passive carrier of foreign particles that could affect the health of an organism." Putrid smells were avoided. Soon the adequate stimulus for olfaction as particles in the air was recognized. It was known that olfactory receptors were in the nose, but the exact locations of the receptor-bearing cells were not known until the end of the 19th Century. As with vision, investigations then proceeded to the central olfactory pathways in the brain.

For motor pathways, introduced in Chapter 3, some of the initial findings reported paralysis on the side of the body that was opposite to brain damage that was limited to the cerebral cortex. Theorists supposed that the motor cortex was toward the front of the brain. But, in Finger's words, the "unequivocal experimental confirmation of a 'motor' cortex" electrically stimulated that part of the cortex and caused movement. Confirming their results, subsequent removal of that part of the cortex of laboratory dogs led to motor deficits. Then neurophysiologists would go on to define "motor cortex" precisely and to describe the motor tracts that lead from the forebrain toward the spinal cord.

Chapters 4 and 5 deal with the neurobiology of emotion and the ability of robots to behave *as though* they had emotion. Early ideas about emotion emphasized our visceral nervous systems, including both the sympathetic nervous system (raising blood pressure, heart rate, etc.) and the parasympathetic nervous system (more or less the opposite effects of the sympathetic). In fact the great psychologist–philosopher William James, at the end of the 19th Century, actually proposed that we feel emotions consequent to changes in the autonomic nervous systems — feelings secondary to vascular changes. At Johns Hopkins, Walter Bradford Cannon and Philip Bard took a more straightforward view, because they were able to stimulate the hypothalamus and directly cause emotional changes in experimental animals, for example, changes like the induction of rage behavior. In subsequent years, the circuitry of the forebrain connected intimately to the hypothalamus (where I have done most of my work) proved to be essential for the performance of all emotional and motivated behaviors.

In the 19th Century, clinicians had to deduce "how the brain works" by observing how behavioral capacities changed after brain damage. The British neurologist John Hughlings Jackson inferred which brain centers were "higher" and which "lower" by carefully noting how certain epileptic seizures in a given patient changed across time.

I think the breakthrough to modern neurobiology occurred when chemical stains were discovered that would reveal microscopic details of nerve cells. The Italian scientist Camillo Golgi got a lucky break when night-time cleaning personnel, servicing the hospital kitchen that Golgi had turned into a laboratory, knocked one of his human brain specimens into a slop bucket. Intrigued by the apparent staining of cells in that specimen, Golgi found that a key ingredient in turning some of the neurons dense-black was (and still is) silver nitrate. The Spanish neuroanatomist Ramon y Cajal was a brilliant exponent of Golgi-stain-based nerve cell biology.

Cajal clearly stated the "neuron doctrine": the brain is not just a continuous string of fibers forming anastomoses (an anastomosis a connection between two structures) to make never-ending nets, but each nerve cell is an autonomous unit. "The neuron is the anatomical and physiological unit of the nervous system." So the necessary question became how neurons were able to communicate. The Nobel Prize–winning physiologist Sir Charles Sherrington (1857–1952) developed the concept of the synapse and introduced modern neurophysiology in his 1932 book, *The Integrative Action of the Nervous System*.

For decades, the development and use of new neuroanatomical techniques dominated the scene. For example, a Dutch neuroanatomist Walle J.H. Nauta (my teacher), who had survived World War II by eating tulip bulbs, came to the States and developed techniques for seeing very fine nerve fibers. This type of technical development led up to the current state, where neuroscientists are trying to map all the connections in the human brain.

After microscopic techniques for looking at neurons gave our field a running start, scientists who were proficient in electrical recording invented what is called "neurophysiology." For example, in Britain, Lord Adrian received the Nobel Prize for showing how to record from individual nerve fibers. Later, tiny wire probes called microelectrodes were developed so that we could put them deep into the brain and record the electrical activity of individual neurons. In addition, of course recording in a noninvasive manner on the surface of the skin over the skull gives you EEG: electroencephalography of wave-like activity of the cerebral cortex, so useful for clinical diagnosis, as in epilepsy or sleep problems. Right now, I use a modern development of the microelectrode: a tiny pipette that suctions onto the surface of an individual neuron, breaks through that membrane, and records from inside the neuron. This is the "patchclamp" technique. It uncovers sophisticated modulations of nerve cell electrical activity that have no analog in robot electronics.

Later still came the origins of neurochemistry; the focus was on how neurotransmitters such as dopamine and acetylcholine are produced in neurons, released at synapses, and eventually broken down. The Nobel Prize winner Julius Axelrod, running a large lab at the National Institutes of Health, became famous not only for his own work but also for mentoring an entire generation of brilliant neurochemists. One of them, Solomon Snyder, not only discovered opiate receptors in the brain but also could claim such a large number of advances that the entire department of neuroscience at Johns Hopkins Medical School is now named after him. As it became easier to study DNA's chemistry and its regulation in gene expression, neuroscientists jumped on the bandwagon. For example, I was able to prove that expression of a particular gene (that which codes for an estrogen receptor) in particular neurons of the brain (hypothalamic and pre-optic neurons) is absolutely essential for specific instinctive behaviors (mating behavior and maternal behavior, respectively). Right now, the focus has shifted to the nuclear proteins that coat DNA in the neuron and regulate gene expression.

As recently as 50 years ago, studies that dealt with complex behaviors such as psychology and personality were dismissed as "soft" for possibly lacking precision. But cognitive science has come a long way. Scientific approaches to animal behaviors — which could tell us a lot about their human counterpart — were split into two parts. One approach, called ethology and most popular in Europe, usually treated the natural behaviors of animals in their natural environments. Ethology was rooted in biology. The other approach, experimental psychology, was more popular in America. It derived from physics and experimental psychological studies would feature well-controlled experiments in the laboratory to answer specific, precisely worded questions or to test formal hypotheses. Both of these approaches could be applied to human subjects. Finally and most famously, the Viennese neurologist Sigmund Freud originated the psychodynamic theory of the human mind and brain — psychoanalysis.

Cognitive neuroscientists often unite these studies of behavior with the various neuroscience methodologies and techniques mentioned above. Historically, brain lesions and their behavioral analyses came earliest. Well known currently, for example, is the patient HM. The Canadian neurosurgeon William Scoville removed most of the hippocampus on both sides of his brain so as to prevent continuing epileptic seizures. The Canadian psychologist Brenda Milner then documented his permanent loss of short-term memory. Other studies emphasized language production, as summarized by the Chatterjee and Coslet book noted below.

Looking back, the first great victory was the observation by the French neurologist Paul Broca that loss of a delimited region on the lower side of the left frontal lobe impaired the production of speech. On the other hand, damage to a cortical area farther posterior, near the juncture of the temporal lobe and the parietal lobe, again on the left side, would impair, in Heidi Roth's words, "the acoustic images of words." Patients with this type of brain damage, studied by the German neurologist Carl Wernicke, could not identify or recognize normal speech. However, these studies were based on small numbers of patients. More had to be studied, brain damage had to be better defined, and the language analyses had to be more sophisticated. But Broca and Wernicke had paved the way.

From there neurologists and neuroscientists went on to initiate the study of all aspects of human behavior. Perceptual abilities, introduced in the next chapter, were particularly easy to study, as were motor abilities, whose pathways are sketched in Chapter 3. Motivation and emotion were soon to follow, as discussed in Chapter 4. My own lab has focused on the most fundamental influence within the brain, a concept I call "generalized brain arousal," essential for initiation of all behaviors. On the other hand, neurologists tend to concentrate on specific disorders, such as epilepsy, autism, loss of memory, and addiction.

Neuroscientific work has reached a level of precision such that our data can often be treated with computations based on applied mathematics and statistics. Computational neuroscience can be divided into two areas: analysis and so-called "modeling." The latter means devising computer programs that are supposed to embody the essential features of some well-chosen groups of neurons in the brain. Both areas of computational neuroscience contribute to the type of artificial intelligence that regulates behaviors by robots.

As stated by Eve Marder, a prominent computational neuroscientist at Brandeis University, "computational models are invaluable and necessary in this task and yield insights that cannot otherwise be obtained. However, building and interpreting good computational models is a substantial challenge, especially so in the era of large datasets." Fitting detailed models to experimental data is difficult and often requires onerous assumptions, while more loosely constrained conceptual models that explore broad hypotheses and principles can yield more useful insights.

Writing in *Neuroscience in the 21st Century*, George Reeke at Rockefeller University, envisions modeling of the brain as an obvious approach to answering questions that we all have about the brain: How are sensations categorized, how are actions selected from a given repertoire, how is "motivation" to be conceived? The availability of computers has allowed academic researchers to construct ever more detailed and complicated models of the brain. Some neural modelers actually try to mimic neurons and neuronal systems faithfully, in detail, while others do not and instead, in Reeke's words, just concentrate on devising "rule-based systems." In all cases, the equations that neuronal modelers use to mimic neurons never match the full sophistication and flexibility of real neurons. Nor are the circuitry properties of the human brain truly realized, even in the best models. Yet many neuroscientists are drawn into modeling, and thus a form of AI, because of the considerable number of free software modeling packages. This implies that progress in neuronal modeling is accelerating. Nevertheless, as Reeke points out, the field is not without its shortcomings. For example, in some cases, the equations representing neurons and their connections are so abstract that they lose the properties of real neural systems. In other cases, neuronal modelers will run large numbers of trials and select some in which their favorite ideas work. That, of course, leads to false conclusions.

The operations of individual nerve cells and individual synapses constitute the irreducible base of neuronal modeling, and have absorbed William Lytton at State University of New York Medical Center. One starts with the nerve cell membrane. The equations that represent the membrane in the model contain the elements of electrical circuit theory: resistors and capacitors. Once those equations and their dynamic changes — when electrical current flows, for example, through sodium channels or calcium channels — are in place, it is possible to start building artificial "circuits." One example would be the modeling of a type of connection (addressed in the next chapter) which involves the passage of sensory information through the thalamus with its subsequent impact on the cerebral cortex. That can be modeled, as Lytton has done, using five types of "neurons" and nine types of connections between neurons.

The main point of this book is to discuss comparisons of human brain and robotic behavioral controls, so as to argue that we understand gross emotional controls so well that we will be able to program robots as though they had emotions. Therefore, I must now provide information about the sensory pathways that feed into emotional mechanisms and the motor systems required for any type of behavior. The next chapter will outline — minus too much technical detail — what we know about two important sensory systems, which will help us to make some generalizations about the relation of sensation to emotional output.

Further Reading

Arkin R (2009) *Governing Lethal Behavior in Autonomous Robots*. CRC Press, Boca Raton, Florida.

Arkin R (1998) *Behavior-Based Robotics*. The MIT Press, Cambridge, MA.

Chatterjee A, Coslett HB (2014) *The Roots of Cognitive Neuroscience*. Oxford University Press, Oxford.

Dertouzos M *et al.* (1974) *Systems, Networks and Computation*. McGraw-Hill, New York.

Dyson G (2012) *Turing's Cathedral*. Pantheon, New York.

Finger S (1994) *Origins of Neuroscience*. Oxford University Press, Oxford.

Hodges A (1983) *Alan Turing: The Enigma*. Princeton University Press, Princeton.

Kruger L, Otis TS (2007) Whither withered Golgi? A retrospective evaluation of reticularist and synaptic constructs. *Brain Res Bull* 72: 201–207.

Kurzweil R (2005) *The Singularity Is Near*. Penguin, New York.

Kurzweil R (1999) *The Age of Spiritual Machines*. Penguin, New York.

Lytton W (2012) In: Pfaff D (ed.), *Neuroscience in the 21st Century*. Springer, Heidelberg.

O'Leary T, Sutton AC, Marder E (2015) Computational models in the age of large datasets. *Curr Opin Neurobiol*. 32C: 87–94.

Reeke G (2012) In: Pfaff D (ed.), *Neuroscience in the 21st Century*. Springer, Heidelberg.

Wilson EG (2006) *The Melancholy Android*. SUNY Press, Albany, New York.

CHAPTER 2

Sentient Machines

Emotional reactions, as highlighted in Chapters 4 and 5, are always reactions *to* something, that is, to sensory stimuli of one sort or another. This chapter will provide a couple of perspectives on sensory physiology (the visual and olfactory) which will illustrate this point. Understanding some elements of sensory neuroscience becomes even more important for our consideration of para-emotional robots when we read Professor Lisa Barret Feldman in *Nature Reviews Neuroscience*. She emphasizes that our brains, far from passively receiving sensory stimuli, act instead as a "predictive organ" that sometimes allows us to perceive things that are not there. Since our emotions shape our expectations of what might be there, the connections between sensation and emotion are clearer than ever.

We appreciate the natural world through our seven or so sensory systems, which enable us to experience our bodies and environment, and help us to control our balance. These are vision, audition, somatosensation (touch), olfaction (smell), taste, proprioception (e.g. muscles stretched, joints moved), and the vestibular system (balance and spatial orientation). How do we employ our senses to "situate" ourselves in the world? It is a great time to explore this issue, since our knowledge of sensory systems is expanding rapidly.

However, there will be immediate complications in trying to relate human sensory systems to robots, because it is impossible to generalize about this relationship. One cannot just say that "humans will do something this way while robots will adapt in that way." In some cases, our neural mechanisms are so complex and subtle that I cannot imagine robots' even beginning to copy human solutions to sensory problems. In other cases, the transition to robots looks easy. Finally, robots will sometimes do a lot better, especially if the human has suffered nerve or brain damage.

In all cases, human sensory systems feed into the central machinery that determines human behavior, including emotionally laden behavior. People get emotional "about" something. I emphasize this because — as Chapter 1 explains, and as Chapters 4 and 5 will argue again — we will have to treat robots *as though* they had emotions.

Scientists have done very well in measuring and charting the perceived magnitude of sensory stimuli (in humans) as a function of the actual physical properties of the stimuli (e.g. loudness and brightness). As Esther Gardner, a neurobiologist at New York University, writes in Kandel's *Principles of Neural Science*, "each physical stimulus is represented in the nervous system by means of a sensory code," an abstract representation of physical reality. The details of those codes determine how we tell different stimuli apart. Almost always those codes comprise specific temporal patterns of action potentials, patterns of "spikes" in time. Besides denoting the onset and duration of a given stimulus, those spikes usually reflect the intensity of the stimulus — the higher the intensity, the greater the number of spikes.

The sensory process starts with receptors (proteins and other chemicals in cells able to respond to external stimuli and transmit a signal to a sensory nerve). As Stewart Hendry observes in the Squire textbook *Fundamental Neuroscience*, receptors that provide information about the external world must be reasonably specific to a narrow range of inputs in order to be useful for telling the organism about its environment. Different sensory systems utilize different receptors. And within each sensory system, only a limited class of stimuli will trigger electrical activity in a given receptor.

In order to provide some perspective on the variety of human sensory systems without slavishly marching through all systems in "textbook fashion," this chapter will simply compare two very different kinds of receptors and their associated sensory systems. *Vision* happens very fast and keeps track of extremely small spatial differences. The stimulus is not chemical. For *olfaction*, the stimulus is essentially and exquisitely chemical. Olfaction is slow and is much less concerned with spatial differences (although smells noticeably decrease at a distance). The pathways used by the two senses differ as well. On the way to the visual cortex, electrical signals conveying visual stimulation pass through the thalamus. Olfactory signals do not. As will become clear, the particularities of these two sensory systems do not suggest that roboticists will try to copy them.

One needs to understand the systems as they appear in the human brain in order to appreciate this claim. Only then is it possible to imagine the kinds of workarounds that robot engineers will attempt.

This chapter, therefore, will discuss how the human brain senses its environment by highlighting two very different systems: vision and olfaction. It will then compare human neuronal mechanisms to some of the mechanisms used in robots, so as to demonstrate how the two types of systems became as different as they are. Finally, we will examine how the kind of artificial intelligence used in robots can be captured to enhance human sensation, especially in patients who have been damaged.

Terry Bossomaier's overview of sensory systems, referenced below, explains that the visual system features the greatest information flow and fastest sampling rates. That to say, visual systems have the capacity to keep track of large numbers of low probability events during thousandths of seconds. In dramatic contrast, olfaction has the least of both. With respect to the ordering of human behavior, Bossomaier sees all of the sensory systems "as both supporting and complementing each other." They "suit the animal (or human) as a package" appropriate to the environment. There is no reason why robot designers could not achieve the same synergies.

Vision

Here is a modest summary of some of the major features of human visual systems.

The physiologist John Nicholls observes that the story of how "neuronal signals are evoked by light to produce our perception of scenes with objects and background, movement, shade and color begins in the retina." To the same effect, Markus Meister and Marc

Tessier-Lavigne note that "the retina is the brain's window on the world."

Writing in my text *Neuroscience in the 21st Century*, Andreas Reichenbach, from the University of Leipzig, reminds us that studies of the retina date back to the early 17th Century. At the time, everybody was puzzled that the retina is actually at the *back* of the eye and that light has to pass through not only the eye but also layers of neurons. However, this counterintuitive setup allows the outer segments of photoreceptive cells, namely rods and cones (the main light sensors, in Reichenbach's phrase), to stick into the crucial photopigments at the back of the eye, pigments whose change in response to a photon of light is crucial for the transduction — the change from one kind of energy to another — into an electrical signal.

As K.-W. Yau has pointed out, the study of the conversion of light as absorbed by the "visual pigments" in the eye to an electrical signal has been going on for more than a century. By now this field has achieved a good level of chemical and electrophysiological understanding. We know that humans have two basic kinds of photoreceptors, both depending, for their light sensitivity, on chemicals located on modified cilia. One kind of photoreceptor (the rod) is extremely sensitive and helps us to see at night. The other kind (the cone) is less sensitive and is useful during the day. In both cases, absorption of light activates a series of closely linked biochemical steps that lead to the flow of electrically charged ions across the cell membrane — hence the electrical signal.

Reichenbach also explains the very confusing business of cone photoreceptor cells — some of them are excited and others are repressed by light — as being due to different kinds of signaling by the neurotransmitter glutamate. While most photoreceptors against a dark background are hyperpolarized by light, there are also bipolar cells in a different retinal layer that can give either hyperpolarization (which inhibits any action potential by increasing the stimulus required) or depolarization (the opposite effect), or even biphasic responses. When the cell becomes hyperpolarized, that is because potassium atoms flow from the inside of the cell to the outside. When the cell becomes depolarized, that is because sodium and — to a lesser extent — calcium ions flow from the outside of the cell to the inside. No one expects robot visual systems to duplicate the subtle interweaving of cellular mechanisms found in the human eye. As Nicholls says of the human eye, "a single quantum of light can give rise to a conscious sensation."

Going back to basics and quoting Terry Bossomaier's summary, "Rods are long and thin and sensitive at low light levels. Cones tend to be fatter, containing more photopigment and having a greater dynamic range."

Opsins, light-sensitive proteins expressed in many cells in the retina, are key to the process of light reception in the retina. This process is called "phototransduction." Struck by a photon of light, part of the opsin protein changes its chemical form, and in doing so starts a cascade of cellular events that results in electrical ion flow and an electrical signal.

Here is a further complexity. Recording from these cone cells with receptors tuned for daylight vision, Yau has reported that these cells can give "strongly *biphasic* flash responses, with an initial hyperpolarization *followed* by a depolarization," but that the exact shape of the response depends on the level of steady light to which the retina has been adapted. It is amazing that this story is still developing despite the fact that retinal responses to light constitute one of the most heavily investigated areas of neuroscience. That is to say, within the past couple of years, cell types in the retina have been revealed as light-sensitive that were not previously recognized as such.

One might think that all the subtleties of the neural processing of visual stimuli come from the operations of the cerebral cortex, so famous for its role in human cognition. However, things get tricky way out in the retina, the several-layered cellular structure in the backs of our eyes. First, the physical chemistry of how a biochemical like rhodopsin receives the energy of light coming through the eye is transformed into the energy of a neuronal action potential, a "spike" that is, a neuronal signal. Second, new cell types in the retina are still being discovered, even though retinal physiology has been studied for decades.

As an example of insights from decades ago, Horace Barlow at the University of Cambridge, faced the problem of how the retina determines the direction in which an object is moving. It is complicated. Having recorded electrical signals from large numbers of direction-selective retinal cells, he proposed a theory in which "directionally selective neurons in the visual system show that neurons do not integrate separately in space and time, but along straight spatiotemporal trajectories." Always, in Barlow's thinking, the key factor was to impose a time delay from the neuron perceiving where the movement started so as to achieve simultaneity with a "downstream perceiving" neuron. Barlow's work has found new life in the papers of Sebastian Seung, a physicist–neuroscientist who started in the neuroscience group at Princeton and is now at MIT.

Among the cell types in the retina studied by Seung is a complex nerve cell called a "starburst amacrine cell." Recordings of this cell have led to the conclusion that such cells can detect the direction of a moving visual stimulus. But how? Seung believes, from studying the cell's structure, that different amounts of delay of the inputs to starbust amacrine cells — a longer delay of the signal coming from the direction in which the visual stimulus is moving — permit these cells to recognize that the source of the stimulus is indeed moving in that direction.

The intricate wiring required for this type of direction-of-motion sensing system discovered by Barlow and elaborated by Seung seems unlikely to show up, as such, in robotic controls. Even though engineers who design robots can build in all kinds of amazing, intricate devices, the capabilities of their workarounds remain to be seen.

Especially important for this book is the manner in which our sensorium affects our emotional lives. If the nitty-gritty mechanisms of sensory perception in robots differ significantly from those of humans, the intimate relations of sensory perception to emotions may also differ significantly. What that means for a robot's "perceptions" of human emotional expressions is not yet clear.

* * * * * * * *

What are the physical limits on our visual sensitivities? Bossomaier's book lists two: "photon noise resulting from the quantum nature of light" and limits on resolution "imposed by diffraction, a property of the wave nature of light." Thus, since light exists in both its particulate nature and wave nature, each set of properties imposes its own set of limitations. These limits will apply equally to robot vision and human vision.

A process known as adaptation serves to afford great sensitivity to small amounts of light against a very dim or dark background, while still allowing visual perception in bright light. Thus, the operations of the visual system include a vast range of conditions. Humans can read in a dimly lit living room. They can also read on the beach! Further, remember that the cells expressing photoreceptors fall into two major groups. Rods are much more sensitive to small amounts of light than cones. But cones not only can operate with higher amounts of background illumination, but also among themselves have different sensitivities to color, thus allowing color vision.

In addition to detecting the presence of light, the receptor cells feature specific "receptive fields," as explained by Clay Reid and Martin Usrey in the Squire textbook. That is to say, a given receptor cell covers only a very small percentage of the entire visual field. Of course, that is how we tell where the visual stimulus is coming from. Then, as you move up the visual pathways from retina to brainstem to visual cortex, the type of visual stimulus represented gets ever more complicated. To quote Nicholls, "information is never, ever passed on unchanged from level to level." Transformations are imposed at every layer of synapses. Different cortical cells might require horizontal bars or vertical bars at the simplest level, passing on to more complex receptive fields: edges, corners, etc. Perhaps the most interesting types of cells to discuss here would be those used in the perception of faces, discussed below.

The low road, from the retina to the lower brainstem. Coming out of the retina, the axons of retinal ganglion cells carry the visual signals to the brain. These signals contribute to two main flows of information.

One route, the more primitive line of information transmission, goes from the retina to part of the midbrain where neurons are poised for visually guided rapid actions. Especially important here are the controls that orient the movements of the head and eyes. For example, those midbrain neurons are the ones that tell a frog exactly where in its visual field the fly is and will go, thus allowing the frog to strike the fly with its tongue even though the fly is a moving target. These neurons are limited to a specific type of visual stimulus — they carry relatively low information content — and their consequences are limited to simple reflex action. As Barry Stein explains in *Neuroscience in the 21st Century*, this part of the midbrain is known as the superior colliculus (from the Latin word for the "little hills" on top of the brainstem). These neurons in the superior colliculus or optic tectum are crucial for the regulation of eye movements and other eye controls.

Within those "little hills" are layers of nerve cells. While some of the layers are purely devoted to visual signaling, the deeper layers also receive auditory and somatosensory (touch) signals. In *Neuroscience in the 21st Century*, Stein summarizes the situation: the maps of space coming from the different sensory modalities are aligned with each other, thus providing a common basis for the initiation of action. Quoting him with added emphasis, "the overlapping nature of the different sensory maps creates a single multisensory map wherein neurons in a given location of the map represent *corresponding* regions of visual, auditory and body space... thus evoking (directed) activity."

The high road, from the retina to the top of the brainstem, namely the thalamus. The other, more recently evolved path followed by the largest numbers of retinal axons go higher — to the top of the brainstem, a particular part of the thalamus, thus to process visual signals and get them ready for the visual cortex. In the words of Usrey, in *Neuroscience in the 21st Century*, "the primary visual cortex receives *streams* of input" from the visual thalamus. Different types of thalamic cells send separate streams of information to the cortex, encoding not only the site of the visual stimulus but also its color. In the cortex, different cells "are tuned to a multitude of stimulus properties," in Usrey's phrase. Thus, to summarize a complex, hierarchical system in just a sentence, primary visual cortical neurons and the cortical neurons to which they project are responsible for our perception of patterned images and their motions. It may be that robotic visual systems will be optimally designed by being split into two systems: one for rapid action control (in the human midbrain, a region known as the superior colliculus) and the other for vision of patterns and colors (the thalamus).

At the cortex, the shape and color of the visual stimulus, its motion and depth, are all processed in different regions. It is easy to imagine how different features of a stimulus are split among different neuronal groups, each with its special discriminating capacities. However, Charles Gilbert, a Rockefeller University researcher with long experience in the analysis of visual perception, also emphasizes something completely different. He has evidence that (in his phrase) "visual perception is a *constructive* process." That is to say, what we see is influenced by what we expect and what we are paying attention to. Thus, in a recent article, he discusses the dynamic nature of connectivity in visual pathways. Amazingly, when animals have been trained to look for a particular visual shape "the shape selectivity of neurons in V1 (the primary visual cortex) changes to a form that approximates the cued shape or a portion of that shape," in his words. As a neuroscientist who works in brain regions distant from the visual cortex, I am really impressed by this finding, because it means that Gilbert's "top–down" controls can influence neural signaling at the earliest stages of cortical processing. This kind of complexity may be hard to build into robots. As mentioned above, if robot designers try to achieve this top–down control in different ways, the consequences of such workarounds will be totally unknown.

In the cortex, two processing pathways have been sharply distinguished. Pathways that proceed forward from the primary visual cortex along the top of the brain, the so-called dorsal visual stream, tell *where* the stimulus is and directs attention to it. A second pathway streams along the side of the brain into the temporal lobe and tells us about *visual objects*, including faces. We will talk more about faces in a while.

Color. Science is still somewhat uncertain as to how color vision works. The very fact that an expert who has studied vision his entire life, Nigel Daw, wrote that "we do not yet understand how the higher levels of the visual system achieve color constancy" tells us how hard the problem is. We should want to know the solution because of the impact of a visually colorful life on our feelings. At lower levels of the visual system, cells with photoreceptors can have specific sensitivities to red, green, or blue. This satisfies Daw's introductory statement that "to see color, one needs at least two classes of photoreceptor with different spectral sensitivities." Simply describing cortical neurons without understanding how their responses got to be that way, Daw knows that there are visual cortex neurons with complex requirements that involve space as well as color. Our likely takeaway is that given engineers' skill in making a wide variety of filters, color vision will be easy to accomplish in robot visual systems.

Visual depth, distance. The possibility of accurate depth perception in robots is uncertain. When Daw lists the cues used in human visual depth perception, they are so numerous that it seems unlikely that engineers will duplicate them in robots. In addition to ocular, physiological cues (like the convergence of the two eyes' gaze), Daw names kinetic cues ("motion parallax"), pictorial cues like perspective, size and texture gradients, and stereopsis (disparity between signals from the two eyes).

Chances for the application of visual neurophysiology to robotic systems seem especially good because of the intensive and brilliant studies of visual systems over many decades. Another benefit of this long history of investigation is that vision studies have compiled an excellent record of coupling well-formulated theories with precise experimentation. That is to say, more than any other field of neuroscience, specific hypotheses about mechanisms of visual functions have been stated and tested head-on. In this sense, even as visual stimuli submit easily to physical measurement, the intellectual side of vision research most resembles the intellectual style of physics. Li Zhaoping's *Understanding Vision* reflects this tendency as he charts progress in vision from behavior to physiology, to models and theory, and then back to behavior. Thrilling to see. In Zhaoping's case, he looks to information theory — measuring how much we can learn from an individual visual coding signal by measuring its improbability — to provide him with the most *efficient* coding system. This kind of thinking has already been applied to robotic sensory systems and will be again in the future.

One practical application of the neuroscience of vision involves eyewitness testimony. Tom Albright, a leader in the neuroscience wing of the National Academy of Sciences, worked with the experienced federal judge Jed Rakoff to produce a National Research Council report entitled "Identifying the Culprit: Assessing Eyewitness Identification." They concluded: "Eyewitnesses play an important role in criminal cases when they can identify culprits. Estimates suggest that tens of thousands of eyewitnesses make identifications in criminal investigations each year. Research on factors that affect the accuracy of eyewitness identification procedures has given us an increasingly clear picture of how identifications are made, and more importantly, *an improved understanding of the principled limits on vision and memory* [emphasis mine] that can lead to failure of identification. Factors such as viewing conditions, duress, *elevated emotions*, and biases influence the visual perception experience." Their report constitutes a rare opportunity to marry the expertise of a leading neuroscientist with the experience of a brilliant judge, educating both law enforcement officials and the public.

Recognizing Faces

One of the most exciting and emotion-provoking events in the human visual system happens when a face is recognized. Obviously, seeing emotion on other faces can be a large part of our own emotional lives. And face recognition will be so very important for robots. How is face recognition achieved?

Charles Gross, a Princeton neurophysiologist, gave one answer as to how the brain might tell that the animal's eyes are looking at a face. Using extremely thin wires that were engineered to pick up electrical signals from individual nerve cells, Gross recorded such signals from a part of the brain known as the inferotemporal cortex (Figure 2.1). Most of us, when asking how faces could be recognized, would imagine that certain neurons would fire off signals when the eyes look at a face and not at any other time. That is exactly what Gross discovered.

Thus, the very first discoveries, in the early 1970s, were made by Gross, who began to find startling properties of cells in an area

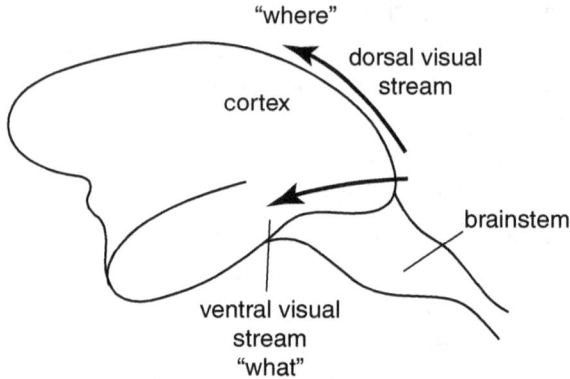

Figure 2.1   Two streams of visual information are sketched: one to the dorsal cortex, encoding "where"; and the other travels more ventrally, to sites like the inferotemporal cortex, encoding "what" (including face recognition).

of the cortex which he had been studying for a very long time, an area which was far removed from the primary visual cortex. Most important, as mentioned, he discovered neurons that would respond only to faces. Some neurons responded best to faces shown in profile; others responded to frontal views of the faces. While these electrical recordings were obtained from nerve cells in monkey brains, it is obvious that similar mechanisms should operate in the human brain and that understanding such cellular mechanisms will prove important for the understanding of certain difficulties of human social behavior. In neuroanatomical work which confirmed and extended Gross' electrophysiology, MIT professor Nancy Kanwisher used magnetic resonance imaging (MRI) to find a convincing "face recognition module" in one part of the human brain. She even identified specific electromagnetic waves associated with the categorization of visual stimuli as faces and the successful recognition of individual faces.

For some time, neuroscientists had supposed that our ability to see other people would depend on what they jokingly called "grandmother cells" — neurons that would fire only if the image of our grandmother (that is, someone who, in general, we would recognize) passed in front of our eyes and therefore allowed us to recognize her. However, we now understand that there are sets of face-selective neurons which are actually members of large ensembles of neurons for recognizing faces. Indeed, Japanese scientists found individual nerve cells that respond only to particular features of the face or head. From their results, we know that facial recognition can depend on patterns of activity among populations of cortical neurons. This is more complex than the idea of "grandmother neurons."

I learned a lot about neural systems for face recognition from approaching my Rockefeller University colleague, neurophysiologist Winrich Freiwald, who records from neurons in the cerebral cortex so as to crack the face recognition problem. I broached the ideas that I am discussing now, and asked for his insights. Specifically, when I asked "what might be the mechanisms that would most easily blend different facial images so that they would seem very similar," he responded: "Don, there are at least two components to your question. One is that face recognition starts with face detection. You've got to know that there is *a face* out there. Second, face categorization, individuation, etc. Interestingly, we found lots of cells in the most anterior group of face recognition cells that *were invariant to identity*. It does still amaze me that this is the case, but it might reflect the need to have a representation of a general face." Thus, Winrich's important answer to the social problem — how images of others could be likened to images of ourselves, for the purpose of guiding social behavior — may have arisen a few meters from my lab: cortical nerve cells whose responses tend to minimize differences among different faces.

Supporting Winrich's work, neuroscientists have recently recorded brain waves from large numbers of patients' brains. They have reported electrical responses to the images of faces that are absent when other visual stimuli, such as strings of letters, are flashed in front of the patients' eyes. Most important, according to these electrical recording results in the human brain, there are "early stages" of face processing and "later stages" of face processing. In this dynamic field of electrophysiology, "late news" will be arriving, detailed and soon.

The plot thickens when we return to still more of the results from Freiwald's laboratory. Among the many fascinating questions he asks, one deals with how the brain can tell one face from another. Indeed, as mentioned, some of the neurons from which Freiwald records signals do respond to a wide variety of faces indiscriminately. I call them "generic" face cells and make special use of them in *The Altruistic Brain*, explaining how we manage to treat others the way we treat ourselves. So, even though generic face cells do not sound very special or impressive, they may be crucial for supporting friendly pro-social behaviors. Predictably, however, Freiwald's work identifies other nerve cells in the cerebral cortex that respond to specific faces.

Using electrical recording, Freiwald, as well as Doris Tsao and Margaret Livingston at Harvard, discovered that almost all (97%) of the visually responsive neurons in a crucial cortical region were strongly face-selective, indicating that a dedicated cortical area exists to support face processing. Further work showed them that several "patches" of face-selective neurons exist and, in their words, so as "to identify the connectivity of these face patches, we used electrical microstimulation combined with simultaneous functional magnetic resonance imaging. Stimulation of each of four targeted face patches produced strong activation, specifically within a subset of the other face patches." As a result, they feel that face patches form a "strongly and specifically interconnected hierarchical network." Some neurons' recognition of a face would depend exactly on what kind of view was presented. Other neurons could recognize a face across a variety of viewing conditions. Thus, this is an exciting current area of research and represents an ongoing quest for understanding "the stability of the single face cell" as it relates both the history of visual learning of each individual and the famously adaptable cerebral cortex.

Just to emphasize, electrical recording is not the only way to discover what is going on with face recognition and, for that matter, recognition of individual properties even beyond faces. Most of the time, when a group of neurons is rapidly firing action potentials, the local blood supply to that exact region of the cortex is increased. Remember that Nancy Kanwisher, the award-winning professor in the Department of Brain and Cognitive Sciences at MIT, used this technique to define the so-called "fusiform face area" (FFA). Subjects in her studies lay very still in a functional magnetic resonance imaging (fMRI) machine and were exposed to a large number of different types of visual stimuli, each for three seconds. She found that the FFA of the human cerebral cortex (Figure 2.1) displayed significant increases in blood flow when the subject was looking at a face but not at other visual stimuli, such as landscapes, objects, or bodies. That is not to say that neurons in the FFA might not respond to other sensory modalities. I once heard Professor Kanwisher suggest that activation there might follow other properties of an individual, such as the voice. But, across all of the cerebral cortex, the FFA gives the most convincing and most selective brain scanning evidence for how we know that we are looking at a face.

As the neuronal processing that results in face recognition goes on, sooner or later the neuronal signals will lead to some behavioral responses. For this book, the most interesting of these are emotional responses. To explain such responses, we turn to the laboratory of Leslie Ungerleider at the National Institute for Mental Health. She found that lesions of a brain region important for emotion, the amygdala, will disrupt the impact of different facial expressions on fMRI responses to those faces that are showing the expressions. Methodologically, combining a brain manipulation with an MRI response to a visual stimulus gave Ungerleider and her team a new approach to facial recognition. As Mark Baxter, of Mount Sinai School of Medicine, observed, the results showed that brain structures "downstream" from the amygdala can respond to amygdaloid damage, which should alter responses to threatening facial images.

What will all of this mean for robotic recognition of faces? The kinds of artificial intelligence (AI) now used for face recognition will surely be built into robots of the future. Among the first scientists/engineers in the field is Joseph Atick, a physicist who started out at Rockefeller University. He left academic life because he saw the importance of automatic face recognition for public security. His insight led to what some call the "biometric security industry." Atick's early achievements were built on straightforward computer programming that led to the automatic recognition of a few faces. The field has exploded from there, academically and commercially. In terms of the applied mathematics, some of the software makes use of statistical comparisons of the features of the person to be recognized with some reference photograph using techniques known as principal component analyses and linear discriminant analyses. Both of these calculations try to take the many features of an image and boil them down to a smaller number of dimensions whose prominence can be measured and compared. These calculations work. For example, recently, the FBI announced the apprehension of a sexual predator and kidnapper by means of this type of technology. And it will get better: three-dimensional imaging now relies on a still more sophisticated program known as electronic bunch graph mapping. All the government needs is an accurate set of reference images of you, me, and the other person, against which to compare images of the person being questioned. Where is the field going? On the one hand, Atick says that, for example, government officials at airports and during traffic stops can help protect the public. Yet the capacity for mass surveillance could rob people of their privacy. As is usual with new technologies, we have the responsibility for maximizing helpful applications and minimizing harm.

The consequences of face recognition are hardly restricted to a cool appreciation of who might be in front of you. There are emotional consequences as well. Consider the work of MIT professor Rosalind Picard, a leader in the famous MIT Media Lab who also started the company Affectiva. An expert in the remote monitoring of autonomic (heart rate, respiration) neural activity, she now utilizes a program called DeepFace, which permits recognition of the emotional components of facial expressions; the program can estimate for one person how another is feeling at any specific moment. A specialist in the technology of physiological recording outside the lab, Picard also has a firm handle on the science, whether focusing on the individual uniqueness of certain emotional patterns or the interactions between the individuals recorded and well-designed robots. A similar program named Affdex can be used for market research, and, exploding out of the MIT academic environment, will probably make money for Affectiva. It seems inevitable that software of this sort will be built into robots in the very near future.

Robotic face recognition may help with this situation: a fascinating application of visual physiology will be to understand what goes wrong in prosopagnosia (face blindness). While we do not know yet what exactly occurs in this disorder, Freiwald has some ideas. In his view, "the best explanation yet (though not fully satisfactory) is that projections from a primary face recognition area into anterior temporal lobe regions" are deficient. In these patients there would be reduced communication between core and extended regions of face processing systems. That is to say, primary responses to facial images by cortical neurons are all right but the integrity of the connectivity in the face recognition cortical network has been damaged. Face recognition by robots, as introduced above, may help patients with this disturbing disorder.

What can we expect from man-made visual devices that can be used with robots? Terry Bossomaier has reminded us that the human eye operates over an extraordinary dynamic range, the ratio of our brightest fields to our dimmest exceeding 10 raised to the 10th power. So there is not much room for improvement in this regard. But Bossomaier also seems to point to game technologies for producing algorithms that will maximize both resolution and speed. In this regard, I will shortly compare human and robotic systems along two major dimensions: constraints on performance and evolution of the systems.

Olfaction

I chose olfaction as the second sensory system because its mechanisms are so different from those involved in vision. Vision is highly spatial and lightning-quick, while olfaction is neither. Further, our touch sensory systems map different regions of the body onto corresponding regions of the brain in a manner that is called "somatotopic." And our auditory systems map the sound frequency range into the brain in a mode that is called "tonotopic." But not olfaction. Olfaction is different.

Kristen Scott, at the University of California, Berkeley, has written about the very large numbers of odors detectable by humans, sometimes in concentrations of less than one part per million. At Rockefeller University, Andreas Keller and Leslie Vosshall have tested human responses to odor mixtures and, following some complicated calculations, have concluded that humans may be able to discriminate as many as a trillion olfactory stimuli. In fact, so far as we know, there do not exist two volatile substances that humans with a good sense of smell cannot discriminate.

Odors are inhaled, and in the nose they must pass through the mucus covering the tiny cilia that stick out from cells that express olfactory receptors. Gordon Shepherd, an expert in olfactory physiology at Yale Medical School, emphasizes that our "wondrously complex nasal cavities" provide a warm and humidified stream of air that eddies across the olfactory epithelium, where the cells with receptors are located. The genes for these receptors, as reviewed by Nobelist Linda Buck and Cornelia Bargmann, encode proteins that are embedded within and on top of the cell surface, beneath the mucus. These receptor-bearing cells express into messenger RNA and protein more than 1000 olfactory genes in mice and, according to Scott, about 500 in humans. When an odor is detected, the receptor-bearing cells respond with an electrical signal (probably because of calcium flooding into the cell) — a signal that is transmitted to the olfactory bulb (Figure 2.2).

In the olfactory bulb, as Scott writes in the Squire textbook, neurons that are expressing different types of odorant receptors send axons to different glomeruli. A glomerulus is a complicated and mysterious microscopic structure in which several cell types are pressed together in a manner that makes them very difficult to analyze. Think of them by analogy to the cartoon in which a football player charges toward a pileup of players and, having disappeared into them, emerges on the other side with his clothing rearranged. At the end of the day, the signal goes to large neurons which are known as mitral cells (shaped like a bishop's mitre) or tufted cells. Altogether, these cells signal in the olfactory code — the result of transformation of odorant chemistry into neuronal firing — a code that gets its specificity from the *combinations* of receptors (and glomeruli) activated as a function of time.

Since even sensory physiology experts like Dale Purves of Duke University, would say that "it is not clear how odor identity and concentration are mapped across an array of olfactory bulb neurons," it

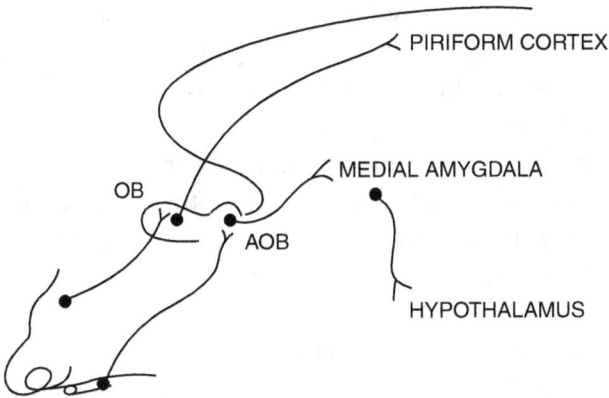

Figure 2.2   Volatile chemicals — chemicals with vapor pressure — are sensed by cells with olfactory receptors in the nose (this is sketched from a drawing purported to be of Leonardo da Vinci's nose). Those cells have axons that project to the olfactory bulb (OB), which in turn projects to the cortex. In contrast, pheromones, with much, much lower vapor pressure, are sensed in a small tubular structure known as the vomeronasal organ, above the roof of the mouth. Those receptor-expressing cells project to the accessory olfactory bulb (AOB), which in turn signals to the medial amygdala and the hypothalamus. Many instinctive behaviors depend on this latter, pheromone-signaling system.

seems likely that human olfactory neurophysiology will be of no use in designing robot systems.

A major problem for us right now in olfactory physiology is that the receptor cells are not constant presences throughout our lives. There is significant replacement of these neurons by continuous differentiation from precursor, non-neuronal cells. How does the olfactory code remain constant and interpretable by the brain when the receptor cells at the front end of the olfactory system constitute an ever-changing population of cells?

Surprisingly, signaling in the olfactory bulb does not depend on odor input alone. Scientists working with Neurophysiology Professor Michael Shipley at the University of Maryland Medical School, discovered that large neurons in the basal forebrain that use the neurotransmitter acetylcholine confer an excitatory bias on olfactory bulb neurons. Likewise, the neurotransmitter serotonin shapes the signaling profile across neuronal populations in the olfactory bulb. Most likely, there is no reason why this kind of central system feedback onto chemical sensors could not be built into robotic systems as well. Some sensors might be amplified in robot systems, while others might be damped — all in the service of producing the desired robotic response.

A major and potentially problematic feature of neuronal responses in the olfactory bulb is their intrinsic cell-to-cell variability in their baseline activity as well as their responses to a given odor. I have recorded that myself when recording electrical signals from a mitral cell and measuring its response to benzaldehyde (smells like the German candy marzipan). I have recorded from the mitral cell "next door," an adjacent neuron in the same animal, and during the same electrophysiological recording session. Indeed, that is a much different level and type of response. Nathan Urban at Carnegie Mellon has tackled this "diversity problem," and has devised an interesting biological rationale. He developed a computational model of olfactory bulb neurons and asked "how effectively different populations of mitral cells, varying in their diversity, encode a common stimulus." Surprisingly, he reported that an *intermediate* defined level of neuronal diversity — not the most similar, not the most divergent — permitted the neurons in his model to achieve "more efficient and more robust encoding of stimuli by the population as a whole." Robot design engineers take note!

Pheromones. Pheromones, briefly, are chemicals secreted by animals that effectively signal their presence to other animals. A parallel system (Figure 2.2) for pheromones also operates in many animal brains and perhaps in some humans as well. Just above the roof of the mouth is a cigar-shaped structure known as the vomeronasal organ, which actually gets its pheromonal input through a tiny tube from the roof of the mouth. The receptor cells in the vomeronasal organ project to a different structure, the *accessory* olfactory bulb, whose neurons in turn project to the medial part of the amygdala. Since this part of the amygdala is deeply concerned with highly emotional matters, like sex, it is easy to see how different pheromones could affect sequences of behaviors in animals and, potentially, emotional states in humans. Some would say that this pheromonal system is most important when babies are still in the womb, but this conjecture is hard to prove.

For the olfactory system, chemicals need to have a vapor pressure — they must be volatile. Pheromones are often much heavier molecules that are deposited by animals on various surfaces, like trees or fences, when the animal rubs against them. Then the recipient animal actually "tastes" them with lips and tongue, and transmits them up through that tube in the roof of the mouth.

The major importance of the reception and processing of pheromones comes from their effect on natural, instinctive behaviors. Most obvious are the sex pheromones, which in a variety of animal species allow animals to enter what I call "hormone-dependent behavioral funnels" that bring reproductively competent males and females to the same place at the same time. But, in fact, behaviors such as fear and aggression are also bound up with specific odors. Most likely, modern chemical-sensing technology will allow specifically sensitive surfaces to be built into robots such that these machines can respond to chemicals with specifically programmed acts.

Molecule by molecule. As with vision, the laboratory of K.-W. Yau has provided some of the most interesting recent reports. Recording from cells in the olfactory epithelium, the Yau team first "obtained a best estimate of the size of the physiological electrical response successfully triggered by a single odorant-binding event." Then, measuring the current produced by that event, they calculated that "it takes about 35 odorant-binding events successfully triggering transduction during a brief odorant pulse in order for an ORN to signal to the brain." In later experiments, Yau's team reduced that estimate to 19 events.

Only 19 molecules of odorant to signal an olfactory sensation to the brain! Do we believe that? Yes, because of previous work with honeybees. German scientists at the Max Planck Institute did electrophysiological recording from the antennae of male honeybees, measuring the responses to the female bee sex pheromone, bombykol. They found that only seven molecules of bombykol arriving at the antenna were sufficient to make the male turn upwind toward the presumed location of a female bee.

One olfactory neuron, one receptor type — this seems to be the rule. In fact, there is some kind of genetic feedback such that once the gene for an olfactory receptor protein has been expressed, others that would have coded for other olfactory receptors are suppressed. The proteins that do the amazing job of olfactory sensitivity are a kind that biochemists and cell biologists know very well. Abbreviated to "GPCRs" for complex biochemical reasons, they are proteins that cross and recross the receptor cell membrane seven times and, once bound to an odor, trigger a well-charted cascade of biochemical events in the cell's cytoplasm.

Although the olfactory system does not appear as orderly as the visual system, which features spatial mapping of the visual field, or the auditory system, which features a tonotopic organization, there are some regularities. Nathan Schoppa's laboratory at the University of Colorado showed that one kind of neuron in the olfactory bulb, the tufted cell, projects only to a basal forebrain group known as the anterior olfactory nucleus, whereas the huge mitral cells project farther, to the piriform cortex. Neuroscientists are now trying to figure out the meaning of this distinction for the regulation of smell-guided behaviors.

What about human pheromones? Much evidence has accumulated to suggest that they do not really exist, but neuroscientist Ivanka Savic-Berglund at the Karolinska institute in Stockholm, has used fMRI to show that brain-imaging responses to sex-related odors are different from each other. Some will say that the vomeronasal system operates in at least some individuals, and a tiny pathway known as the nervus terminalis could serve sexual responses at the behavioral level. Obviously, none of this is going to happen in robots.

Perfumes. Some people disparage the importance of smells in human life in comparison with animals like beagles or raccoons, which are famous for having important behavioral decisions depend on their olfactory sensitivity. Such people conclude that smells are not important in human life. Not true. For humans, smell does not count? Tell that to Givadeaun, Inc., or to International Flavors and Fragrances, Inc., or any of the other companies in a $39 billion fragrance industry.

Deodorants. Likewise, for those who have not been on a crowded New York City subway recently or, perhaps, in an athletic locker room, it may be hard to understand that covering up unpleasant body odors matters to millions of people of both sexes at every socioeconomic level. The deodorant industry is worth more than $16 billion annually just in the US.

Flavors. One of the more important applications of our knowledge of the olfactory system in human life depends on the contributions of smell to the flavors of food. It is widely understood that the flavors of foods depend on a complex mixture of taste and olfaction. Not surprisingly, the subject has a history.

In the 18th Century, it was the servants — not the family upstairs — that got to enjoy food intensely. This was because the servants occupied confined, often poorly ventilated kitchens where the aromas of roasting meats and boiling soups would concentrate for hours. The people who prepared food could taste food as it cooked, literally surrounded by wafting scents of spices and large joints of beef basted in butter and orange juice. In the 18th Century, there were no kitchen stoves; food was exposed on rotating spits and in open cauldrons hung over the fire. For the servants, food was first an olfactory experience: when they sat down to eat at the kitchen table it tasted better than what arrived upstairs after riding the dumb waiter. Distance made food less tasty, and the reason was that its aromas dissipated. Today, there is a reason why we like to eat in the kitchen. To the extent that we can still smell the food as it cooks, we can anticipate how good it will taste. The scent makes us think about the food and taste confirms what we are thinking.

One reason that we like to cook when we entertain, rather than just ordering elegant takeout, is that the aromas of food tend to bring people together. Aromas are shared, creating a sort of primitive collective experience that makes even unacquainted people feel like a group. Inviting people half an hour before the roast comes out, then letting it sit for ten minutes before the carving, has a noticeable effect on conviviality. When people do finally eat, their expectations have been raised and they almost automatically like what is served. This is why many cookbooks show people standing around in large suburban kitchens, not only nibbling *hors d'oeuvres* but also taking in the scent of whatever is cooking. This is also why fragrant stir-fries entice everyone into the kitchen if they have been lingering somewhere else. The olfactory element of food acts as a promise — indeed, an inducement which can be raised or lowered depending on what we cook. Heady spices will increase the inducement, while bland scrambled eggs will dampen it. Ever seen how cookbooks tell you to roast your spices before you grind them? You are releasing oils that literally go to people's heads before they even take a bite. Once they do, the spices are that much spicier.

When we remember food, we often think of how we breathed it in before we even tasted it. Baked rice pudding is a perfect example. The vanilla filled the kitchen. The cinnamon on top drove us wild. Stove-top rice pudding, flavored with cardamom pods, was a show-stopper hours before the final product came out of the refrigerator with those pods neatly removed. Now, think of cooking steak over a fire when you were at camp. The steak might have been tough and chewy, but what you remember — and would die to recover — was how the smoke hit your nostrils and made you ravenous. Again, it was the immediacy of contact with food — the actual baking, boiling, and broiling — which is often defined in olfactory terms. Such immediacy is now missing from our microwave culinary culture and we often eat more just to compensate.

In many of the best restaurants, the kitchen is as far as possible from the patrons. Your first experience of the food is when a plate is set before you. But think about some of those old ethnic restaurants, like maybe the Italian mom-and-pop places you still occasionally find. No professional restaurant designer decided where to locate the kitchen. Instead, the kitchen (invariably) sits behind a swinging door just behind the back tables. Every time a waiter emerges, you can smell the garlic. It is overpowering. You want that food, and you know how it will taste before you order. The smell carries associations — not just of certain kinds of food, but of whole matrices of experience that are now increasingly hard to come by. The smell brings it all back. There is a direct connection between what is lodged in the memory and what is on the menu.

At a very deep level, food is not just about taste, but about how it situates us both in a group and relative to our personal history.

Smell plays an important part in stimulating these connections. Many women who could buy the best of everything still bake bread because they want their children to develop the same sort of associations that they had growing up. Bread fresh from the oven filled the house with incomparable aromas that signified love, family, a certain skill with yeast and flour. These signifiers are worth passing on.

Of course, there are many of us who do not think about food in its broad context, but just want it to taste good. When we eat a salad, we like the crunch of the vegetables, but we also take in the pungency of oil and vinegar dressing. We like that food can be experienced as being complex, carrying olfactory sensations as well as taste and feel. When we have a cold, and are deprived of being able to smell our food, even our favorite dishes become suddenly boring. It is as if the fall-off in smell inflected all of the other ways that we experience food. In *Season to Taste: How I Lost My Sense of Smell and Found My Way*, Molly Birnbaum describes the excruciating experience of being passionate about food (she had a scholarship to the Culinary Institute of America) and then suddenly losing her sense of smell. Ultimately, she recovered and pursued her passion. But the agony of olfactory deprivation was almost unbearable. Smell is as intrinsic to our experience of food as any other sensation, and without it the others almost cease to matter.

But, even beyond food, there is wine. No wine connoisseur would ever think of taking a drink before he first sniffed his glass. Sniffing becomes part of the ceremony of celebrating a wine. It is a way of setting up the wine, as if failing to first sniff it would somehow be unfair to a beautifully aged product of terroir and skill. While such sniffing cannot be abstracted from ultimately drinking and tasting, there is no way of imagining consumption of the wine without experiencing its aroma. In this sense (no pun intended), wine provides perhaps the essential example of how smell matters to human ingestion.

On a more mundane level, think of the last time that you ran through an airport and almost stopped for a snack that you did not need. Chances are it was because the Cinnabon kiosk was pumping out a heady cinnamon scent impossible to resist. Smelling food makes us want it, and it may be nature's way of ensuring that we eat when we have the chance even if we are not starving.

The cooperation of taste and olfaction to make flavors represents just one example of what Lawrence Rosenblum, Professor of Psychology at the University of California, Riverside, calls "a confederacy of senses." Our ability to understand speech is improved by being able to see the lips of the speaker. Different sources of information reinforce each other; for example, the manner of speech tells you who the speaker is. In Rosenblum's words, "The multisensory revolution has already started to help people who have lost one of their primary senses." In terms of neuroscience, the new perspective includes the recognition that the cerebral cortex is not just sliced up neatly into different sensory zones, but that some cortical areas clearly are responsible for combined signaling from different sensory modalities. The resulting sophistication of multisensory signaling may be hard for robot engineers to achieve, at least in the near future, because neuroscientists are not yet even sure how Rosenblum's confederacy of senses works.

Olfactory receptors outside the nose. Hans Hatt at Uhr University in Germany, has discovered that our skin cells have olfactory receptors on them. Neuroscientists do not really know why. While olfactory receptors had already been found in skeletal muscles, their importance also remains to be discovered. Presumably they have something to do with biochemical sensing and cellular growth, but whatever the reason they are not likely to find their equivalent in robots.

Overall

In sum, as Peggy Mason's text points out (not just with respect to vision and olfaction but for all of our senses), while there are a limited number of receptor types, "the world contains a continuum of stimuli with an infinite number of stimulus properties." What Mason and other neuroscientists would always say is that *combinatorial codes* based on the outputs of our sensory receptors allow our brain to appreciate and discriminate a much larger range of stimuli than would otherwise be possible. There is no reason in principle that robot designers cannot achieve the same types of combinatorial codes and thus increase the range of stimuli that can be sensed and responded to.

Comparing Humans and Robots

No one expects robots to solve every behavior control problem in ways that humans do. In fact, as noted below, when they don't, humans and robots will have especially good opportunities to help each other.

After all, look at the nerve cell units that human brains possess. Figure 2.3 depicts two very different types of neurons. One, typical of sensory signaling pathways is called "allodendritic" and features

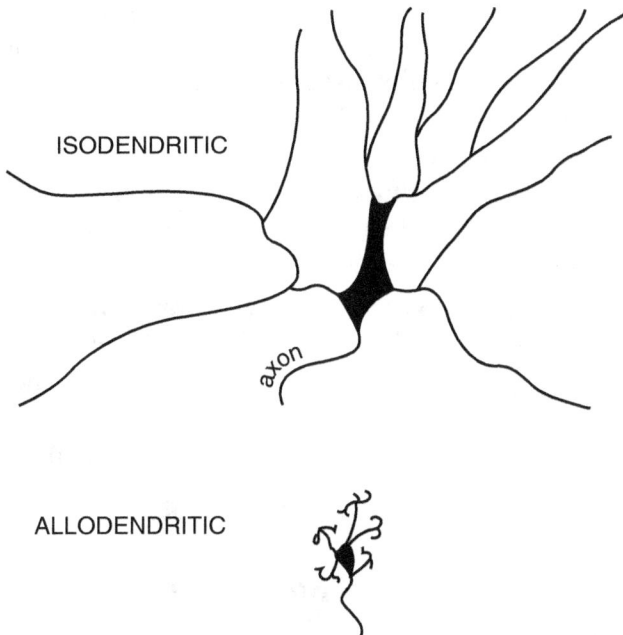

ISODENDRITIC

axon

ALLODENDRITIC

Figure 2.3 Two complex patterns of elaboration of neuronal dendrites are sketched here and contrasted. "Isodendritic" patterns have each additional dendritic segment (as you travel out from the neuron's cell body, here in solid black) *longer* than the previous segment. This pattern allows for an incredibly wide catchment area of information signaling to this neuron (redrawn from an actual reticular neuron studied by M. Serafin in the Physiology Department at CMU in Geneva). It is more typical of arousal systems and emotion systems in the brain, and shows up in our Figures 3.2 and 4.4. "Allodendritic" patterns have the opposite: each additional dendritic segment is *shorter* than the previous segment. It is more typical of specific sensory systems, such as certain parts of visual and auditory systems.

Table 2.1   Comparison of Human and Robot Motor Controls

	Human	Robots
Evolution	Biological	Industrial design
Constraints	Material strength, speed	Economic, marketplace

small, dense, bushy dendritic trees. The other, typical of brainstem arousal pathways is the opposite and is called "isodendritic." The dendritic trees stretch out widely. The former permits precise signaling in a narrow sensory range. The latter permits broad integration among different signals. So far, chips used in robots — even including the chips heralded recently by IBM — do not match the sophistication of what we have in our heads.

Our visual systems can function on a sunny beach or in a dark room. We begin to sharpen images starting immediately in the retina. There are no two volatile compounds that the human nose cannot distinguish. So, even as we envision cooperation by humans with robots, human sensory systems in their own right are nevertheless extraordinary and virtually certain to remain different from their robotic counterparts.

I envision major sets of differences between human brains and robotic systems in the language of Table 2.1.

Human Constraints are Not Robotic Constraints, and Vice Versa

Everybody has problems, and that sentiment certainly applies to the performance of behavioral responses by humans and robots. In fact, comparing the problems to be solved in human and robot sensation offers us one of the interesting ways of comparing the two.

Beginning at the beginning, exactly how are robot control system engineers going to wire things up? If those engineers are mimicking the brain, then the chemistry is tricky, the spatial precision required is on the order of thousandths of a millimeter. Beyond that, can we find the right metallic materials (some of which come from obscure places in Africa) and can we limit costs well enough to produce a profit and satisfy activist shareholders?

However, the robot will indeed be able to pick up electromagnetic vibrations outside the visual range, and will be able to make the finest discriminations, for example, among shades of a color. These will much exceed what even the most sensitive humans can achieve. Robots might be able to help with problems of face recognition, for example, with patients who have the failure of face recognition in the disease known as prosopagnosia. However, some of the cortical neurons recorded by Winrich Freiwald may be so subtle in displaying their properties that we do not quite know what purposes they serve and therefore could not mimic them in robot behavioral control systems.

As to chemical sensing, it is clear that during manufacturing processes, for example, robot sensors can detect various kinds of imperfections in the half-finished products that humans could not see or feel.

Evolutionary Mechanisms

Likewise, human sensory systems evolved by different mechanisms than robotic systems. We evolved through the processes of mutations in DNA followed by natural selection and survival of the fittest in a biosocial world. By way of contrast, today's robotic systems were developed by engineers skilled in the commercial field of industrial design. Much different.

It is crucial to understand that precisely because the constraints and evolutionary backgrounds of human brains and robotic control systems can complement each other, they are ideally suited for synergy. We will return to this in the final chapter.

Aiding Humans with Damaged Sensation

Neuroscientists instinctively ask: "How can my discovery feed into some kind of therapy?" This book extends that question to ask: "How can the types of artificial intelligence applicable to robots be used to help humans?"

Simply stimulating the visual cortex to mimic normal pattern vision did not work very well. But look to the retina. Injecting electrical impulses into the output cells of the retina, the neurons that signal to the brain, may work. Here is a glimpse of two scientists leading this quickly emerging field of work that combines neurophysiology, electrical engineering, and computer science.

Consider patients with damaged retinas. Eberhardt Zrenner, in the Center for Ophthalmology at the University of Tuebingen, Germany, began years ago, to develop a device, Retina Implant AG's Alpha IMS, for stimulating retinal neurons in patients with the disease retinitis pigmentosa. He thinks that the disease "results in a complex self-sustained outer retinal oscillatory network" that is abnormal. To correct that, he considers several technical options, including exactly where to put the implant for the purpose of stimulating electrical activity. Even in a place as small as the back of the eye, there are some choices to make. His other implant, Argus II, is placed on the surface of the retina, and he has reported that patients did "statistically better with the device switched on versus off in the following tasks: object localization, motion discrimination, and discrimination of oriented gratings." Back to Alpha IMS — he reported: "The subretinal implant restored light perception in 8 out of 9 patients (8/9), light localization (7/9), motion detection (5/9; angular speed up to 35 degrees/second), grating acuity (6/9), and visual acuity measured with Landolt C-rings (2/9) up to a visual acuity of 20/546. Identification, localization, and discrimination of objects improved in repeated tests over a 9-month period." He wants to do better. For example, spatial resolution, temporal resolution, and color vision all provide challenges for the future.

Sheila Nirenberg goes deeper. A MacArthur "Genius" Award winner and a professor at Weill Cornell Medical College, she wants to make successful retinal implants but knows that to accomplish such a goal to her satisfaction, she must understand how retinal signaling works. She states that retinal prosthetics to date have not worked very well because they are not duplicating normal retinal signaling. Thus, for years, she has been working with fellow Cornell professor Jonathan Victor, a medical doctor and a leading computational neuroscientist, to figure out exactly which features of the time series of neuronal action potentials coming out of the retina are most important for veridical sensory signaling.

They have devised an exciting possibility — a mathematical model of retinal signaling that builds on "principles of linear–nonlinear cascade models, but stands apart from other approaches in that it generalizes to stimuli of essentially arbitrary complexity, that is, it captures the transformation from visual input to retinal output for essentially any image in real time." Even two years ago, Nirenberg reported that by using signaling strategies that try to incorporate the retinal neural code, she could come closer to normal image representation. Of course, her work has the greatest potential not only for patients but also for the design of robotic visual systems.

Other possibilities include gene therapy and cell-based therapy. Some scientists have used adenoassociated viral (AAV) vectors for gene therapy. They inserted so-called "optogenetic molecules" into the retina in order to render retinal cells light-sensitive. Other molecular biologists experiment with different types of viral vectors and in doing so warn that difference among species (e.g. mouse, cat, human) might affect the efficiency of such gene therapy.

As I see it, robotic systems will be able to take every advantage of the accomplishments of scientists like Nirenberg, whereas all the cell biological advantages will be quite irrelevant to robot futures.

It has surprised me greatly that the kind of artificial intelligence that has fed the "explosive growth of the graphics and sounds used in computer games," in Bossomaier's phrase, has, *pari passu,* led to the development of devices that could help damaged human patients. In turn, he says that the industry has moved from purely deterministic games, like chess, toward virtual worlds that are right up the robot designer's alley.

Moving toward the intellectual considerations of sensory perception in strictly quantitative terms, performances in robotic sensory systems can be measured in the units of information theory. How much information is in a stimulus set? That set is measured proportionally to the logarithm of the number of possible stimuli. Can the robot system handle a given set? How does the robot system perform in the presence of noise, or static, or whatever distractions are there? Ultimately, in terms of neuroscience "decision theory," robot engineers will require high sensitivities of sensory systems yielding large numbers of correct choices, coupled with low numbers of erroneous choices.

Outlook

Things in this field are moving fast, right now. Neurophysiologists and cell biologists, and even stem cell experts are deepening and expanding our knowledge of visual systems daily. At the same time, scientists and engineers sense that commercial possibilities of exploiting potential robotic capacities are on the horizon. At the best institutions here in the States and abroad, the understanding and improvement of our sensory systems constitutes a hot topic.

For example, the improvement of our visual system: seeking scale, improving sensory sharpening capacities, and allowing for technical evolution to get even better. Ramesh Raskar at MIT is doing his best, trying for "superhuman vision," trying to "make the invisible visible." Right now he has new software for reducing the blur in moving images. Superfast imaging processes could easily be built into robot vision. Likewise, Gershon Dublon and Noseph Paradiso of the MediaLab, also at MIT, have remarked on the tremendous numbers of electronic sensors in our environment that are, as well, network connected. As they said, "Cameras and microphones in your computer, GPS sensors and gyroscopes in your smartphone, accelerometers in your fitness tracker," and, in your new home or office, "sensors that measure motion, temperature and humidity." With each decade, their sizes decrease, even as their sensitivities and signaling capacities increase. Obviously, all of this could be applied to robot sensation. A person or a robot wearing an electronically identifiable ID tag can thus make available to that person *or robot* a tremendous amount of sensory data about his or her environment.

In spite of the contrasts between human sensory systems and robot sensory abilities (for example, in Table 2.1, robot engineers clearly will get robotic sensory systems to do a large percentage of what humans can do, and more. Nirenberg told me that robots might not use the same cellular structure as, for example, the human retina, but that robot engineers and scientists [like herself] who design sensory prostheses will get the job done in other ways.

Further Reading

Bhandawat V, Reisert J, Yau KW (2010) Signaling by olfactory receptor neurons near threshold. *Proc Natl Acad Sci USA* 26; 107(43): 18682–18687.

Bossomaier T (2012) *Introduction to the Senses*. Cambridge University Press, Cambridge.

Cao LH, Luo DG, Yau KW (2014) Light responses of primate and other mammalian cones. *Proc Natl Acad Sci USA* 111(7): 2752–2757.

Daw N (2012). *How Vision Works*. Oxford University Press, Oxford.

Freiwald WA, Tsao DY (2010) Functional compartmentalization and viewpoint generalization within the macaque face-processing system. *Science* 330(6005): 845–851.

Kandel ER *et al.* (eds.) (2013) *Principles of Neural Science*, 5th edn. McGraw-Hill, New York.

Luo DG, Xue T, Yau KW (2008) How vision begins: an odyssey. *Proc Natl Acad Sci USA* 105(29): 9855–9862.

Mason P (2011) *Medical Neurobiology*. Oxford University Press, Oxford.

Moeller S, Freiwald WA, Tsao DY (2008) Patches with links: a unified system for processing faces in the macaque temporal lobe. *Science* 320(5881): 1355–1359.

Nicholls J *et al.* (eds.) (2001) *From Neuron to Brain*, 4th edn. Sinauer, Sunderland, MA.

Nirenberg S, Pandarinath C (2012) Retinal prosthetic strategy with the capacity to restore normal vision. *Proc Natl Acad Sci USA* 109(37): 15012–15017.

Pfaff D (ed.) (2012) *Neuroscience in the 21st Century*. Springer, Heidelberg, New York.

Pfaff D (2014) *The Altruistic Brain*. Oxford University Press, New York.

Purves D *et al.* (eds.) (2012) *Neuroscience* 5th edn. Sinauer, Sunderland MA.

Sherman S (2005) *Fresh from the Past: Recipes and Revelations from Moll Flanders' Kitchen*. Taylor Trade Publishing, Lanham, MD.

Sherman S (2010) *Invention of the Modern Cookbook*. ABC Clio, Santa Barbara.

Squire L *et al.* (eds.) (2013) *Fundamental Neuroscience*, 4th edn. Academic Press/Elsevier, San Diego.

Tsao DY, Freiwald WA, Tootell RB, Livingstone MS (2006) A cortical region consisting entirely of face-selective cells. *Science* 311(5761): 670–674.

Li ZP (2014) *Understanding Vision: Theory, Models, Data*. Oxford University Press, Oxford.

CHAPTER 3

Motors to Go

In the previous chapter, I sketched some of the sensory capacities of the human brain and compared them to robots. Yet, as must be apparent, any such "comparison" is a difficult, imperfect undertaking, since at this point of our experience with robots we face a fundamental challenge: To what extent do we model them on humans? We sometimes understand robots' brains best by *referring* to humans' brains as a baseline for discussion. Therefore, in this chapter, I will lay out, for the nonscientist, some of what neuroscientists know about how humans control movement.

Our understanding of motor controls from the spinal cord, through the brainstem, right up through the cerebral cortex is expanding rapidly and on many fronts. Here are the basics.

I will use the human brain as a sort of baseline from which robots' brains depart in varying degrees. Since (as I will discuss) motion is necessary for expressing emotion, I will contrast principles of operation of motor mechanisms in robots with the regulation of human head and limb motion by the cerebral cortex, cerebellum, and brainstem. In particular, I will demonstrate that *constraints* on human motor performance are functionally and substantially different from those which limit robots' activities. Limits matter as much as motivation since they restrict, and thus define, *how* we move.

Moreover, given the complexity and high level of integration characteristic of human nerve cells and neural networks, I will argue that it will be difficult, if not impossible, to duplicate these traits in robots. In other words, I want to illustrate what we cannot do with robots — at least given our present technical competence — and what we can. Because humans and robots will have markedly different capacities of motor control, they should be synergistic. That is to say, some things that we do worse, robots might do well, and vice versa, so that together we could raise the general level of performance.

I will begin by setting out some of what we know about the brain's capacity to produce and regulate motor activities, naturally including all that underlie emotional expression. I do this because, while the deepest part of my argument in Chapters 4 and 5 will emerge in the chapters dealing directly with the production and regulation of emotion, it is necessary to realize that all emotional expression requires movement. Whether we are trembling with fear or grinning happily, the central nervous system is causing muscles to move. The question is: How does the brain control such movement? How does it control voluntary and involuntary movements, both of which are involved with emotion albeit at different levels? Movement is important because behavior is movement.

We breathe, look, smile, grasp, walk, and run. We make voluntary movements toward objects and toward each other. How do we do these things? I will begin by sketching the lowest levels of movement control in the spinal cord; spinal cord neurons are in turn regulated by signals descending from the brainstem and from the cerebral cortex. I will cite ideas from around the world as to how the human brain produces two very different kinds of movements: stereotyped repetitive movements and newly learned movements.

I will then compare human neuronal mechanisms to some mechanisms used to generate and control robotic movements. There are major differences. But at the end of the chapter I will give examples of how artificial intelligence that controls robotic movements can be engineered to help badly damaged humans.

Movement Controls, Starting at the Spinal Cord

To quote the biophysicist John Nicholls, "the neural organization of motor control is hierarchical, with smaller, simpler elements integrated into more complex patterns at higher levels of the nervous system." Gordon Shepherd, Professor at the Yale School of Medicine, also emphasizes hierarchies of control in motor command systems. According to him, the concept of motor control hierarchies originated with the nineteenth century neurologist John Hughlings Jackson, who studied patterns of movement in certain kinds of epileptic seizures that revealed such hierarchies. Jackson said that "higher levels" of movement would be more recently evolved purposive movements (for example, precisely putting broken eyeglasses together or moving a chess piece), while "lower levels" of movement would include automatic reflex movements (for example, breathing or withdrawing a burnt hand from a fire). During "Jacksonian seizures," higher level movements could recruit lower levels, but not the reverse.

Every overt behavioral response requires muscular contraction. However, under normal circumstances, muscles do not contract by themselves. Their movements must be triggered by inputs from the central nervous system. For most of the body, these inputs come from large neurons with long axons termed "motoneurons" (motor neurons). Leading scholar Robert Burke observed that motoneurons have come under intensive study "because of their critical roles in the control of all movement." Motor neurons connect to specific muscle fibers and regulate their contraction. The size of the motoneuron affects its excitability. Harvard Medical School professor Elwood Henneman enunciated the "size principle": a given amount of electrical current input makes a big difference in a small cell's membrane voltage potential, and that makes for firing early. The same amount of electrical current input to a large motoneuron makes a smaller difference in the membrane potential, and that neuron fires later. But when it does fire, a bigger neuron acts as a more intense, more powerful trigger to movement.

Thus, an important reason for understanding exactly how motor neurons are recruited into the pool of neurons governing a particular behavior is that this will determine how much muscle force will be exerted during that behavior. The more neurons, the more force. And the last motor neurons recruited will provide the largest forces, restricted therefore to behaviors that need that force.

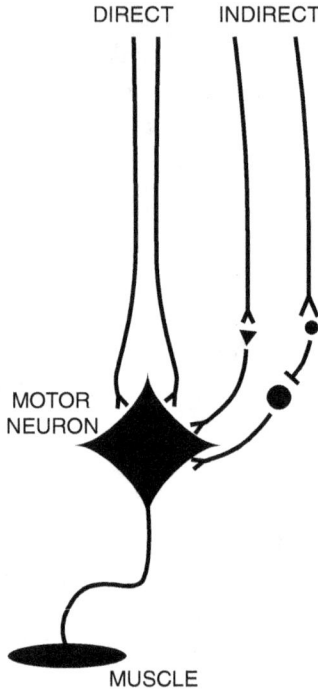

Figure 3.1 Brainstem systems projecting to the spinal cord use both direct connections to motor neurons and indirect connections (through interneurons).

So where do these electrical inputs to motoneurons come from? The brain's most direct controls over movement come from axonal projections directly to motoneurons in the spinal cord. The major alternatives come from the brain's projections to smaller cells in the spinal cord: the so-called "interneurons (Figure 3.1). Some of these projections from the brain are very specific — for example, axons that descended all the way from the cerebral cortex to the spinal cord. These direct connections from cortex to motor neurons are built in only for the most skillful muscle use — for example, the movements of fingers in humans. Such projections can account for the great precision of many of our movements, such as those of a pianist or a painter. Other projections are nonspecific, for example, from the very core of our brainstem, known as reticulospinal neurons. Such neurons can project to all levels of the spinal cord, on both sides (Figure 3.2).

BASAL FOREBRAIN
(bnst, poa, sep)

MIDBRAIN

periaqueductal gray

midbrain reticular formation

Edinger-Westphal
nucleus

deep and intermediate
layers of superior colliculus

oculomotor
nucleus

trochlear nucleus
(small projection)

DIENCEPHALON

medial thalamus

medial and lateral
hypothalamus

PONS

pontine reticular formation

trigeminal nuclei

motor nucleus V

vestibular nuclei

locus coeruleus

raphe nuclei

facial nucleus

NGC

all
levels
of
spinal
cord,
bilateral

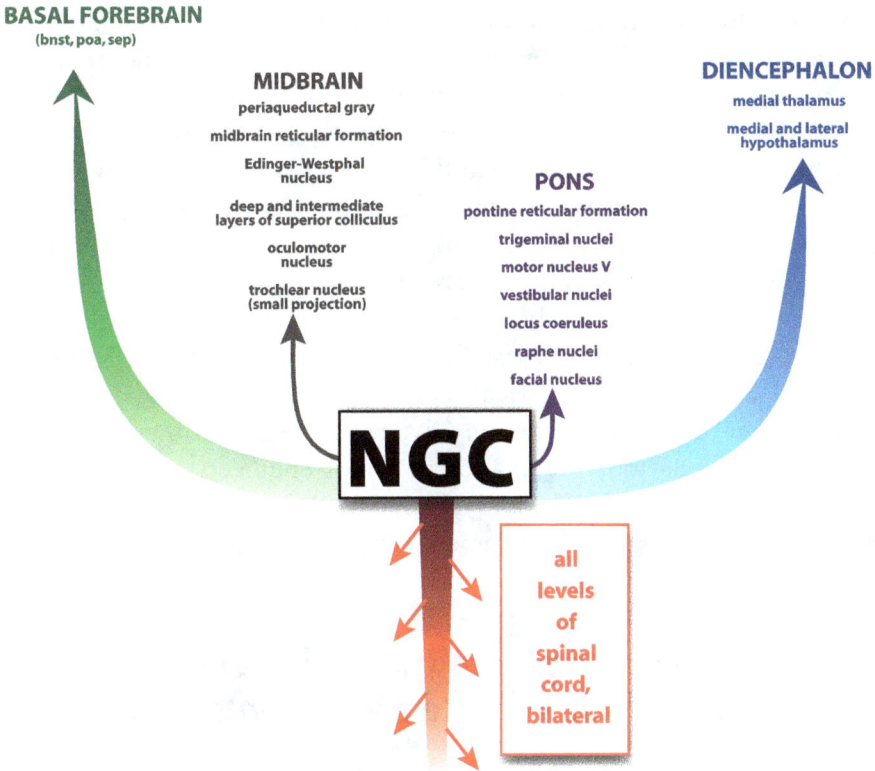

Figure 3.2   In the reticular nucleus gigantocellularis (NGC) of the lower brainstem, very large reticular neurons not only have ascending arousal-related projections, but also its reticulospinal neurons project to all levels on both sides. [Adapted from Pfaff D *et al.* (2012), with permission].

These projections from the ancient brainstem reticular formation to recipient neurons all over the spinal cord maintain the general muscle tone and are necessary for starting just about any movement.

For our arms, legs, and trunks, everything depends on controlling neurons in the spinal cord, which are themselves complicated. Consider that the motor neurons, the very nerve cells that control muscles, are triggered by direct inputs from sensory stimuli, but also from other, unspecified neurons in between — interneurons — and even by the outputs of other motor neurons. In turn, nerve cell pathways coming from the brainstem may act by modulating the signals from those sensory stimuli or by either modulating the interneurons or impacting the motor neurons themselves.

I do not know of any robotic circuitry that is organized this way, in which complex connectivity and logic appear in the vicinity of the final element (in our case, the motor neuron) that will trigger muscle contraction. Indeed, developmental biologist Tom Jessell at Columbia has shown the subtlety and sophistication of the chemical signals that govern the positioning and connectivity of spinal neurons regulating motion. Likewise, Physiologist Elzbieta Jankowska at the Karolinska Institutet has used electrical recording to chart the logical structure of neuronal signaling there. Given that degree of human spinal neuronal articulation, the chance of robot engineers duplicating it is approximately zero. Certainly, comparing human strategies with robot design strategies of movement regulation will be interesting.

Signals Descending from the Brainstem Smoothen and Coordinate Movements

Humans generate behavioral actions in many ways. Biological evolution has arranged for different solutions to different motor control problems. As a result, in the human brain and spinal cord, there is an overlap among different neuronal systems for controlling behavior. This "holistic" approach to motor control distinguishes the human situation sharply from robot engineering, which is very unlikely to be able to imitate this finely orchestrated set of motor controls in our brains.

Movements of the face and eyes are just as complicated, but rather than being controlled from the spinal cord they depend on specialized nerve cell groups in the brainstem. Brainstem triggering and coordination of behavioral movements work both by long-wired systems and by doing "local business" (signaling only with neurons directly adjacent). Of course, in the specialized nerve cell groups for muscles of the face and eyes, there are tremendous numbers of calculations by neurons next to each other signaling rapidly.

We can understand the multiplicity of neurons in the human brain-stem that contribute to movement control by acknowledging that even merely standing up cannot be taken for granted. Maintaining posture depends on several sources of inputs from the vestibular system, the cerebellum, feedback from the muscles of the trunk and leg, as well as powerful outputs from large neurons in the core of the brainstem (the reticular formation).

Just a moment's reflection reveals that all of these systems project-ing from various parts of the brain to the spinal cord work together very well. Otherwise, all of our movements would be random, jerky, and ill-coordinated. For a snapshot of the complexity without going into excruciating details, Figure 3.3 (from Chris Miall at the Univer-sity of Birmingham, England) shows a top-view sketch of the human brainstem naming 17 different sources of axons descending to the spinal cord. These various sources of excitation follow different paths to the spinal cord, with different velocities of spike conduction and patterns of termination in the cord.

According to British professor Roger Lemon, "Each of the descending pathways involved in motor control has a number of anatomical, molecular and neuroinformatic characteristics. They are differentially involved in motor control, a process that results from operations involving the entire motor network rather than the brain commanding the spinal cord." A given pathway can have many func-tional roles.

Following from Lemon's overall view and according to Marc Schieber of the University of Rochester Medical School, motor control systems descending from the brain can be divided into two groups. The axonal systems that run close to the midline of the brain, the "medial systems," are devoted especially to the control of posture. These medial postural systems signal from our vestibular organs the angles and tilts of our heads, and are especially adapted for very fast behaviors that restore our balance. Maintaining posture is complex and difficult, and depends on the terrain, on a wide variety of limb and axial muscles, and on feedback from sensory receptors.

In contrast, according to Schieber, descending axonal systems that are more lateral — farther from the midline — are specialized for

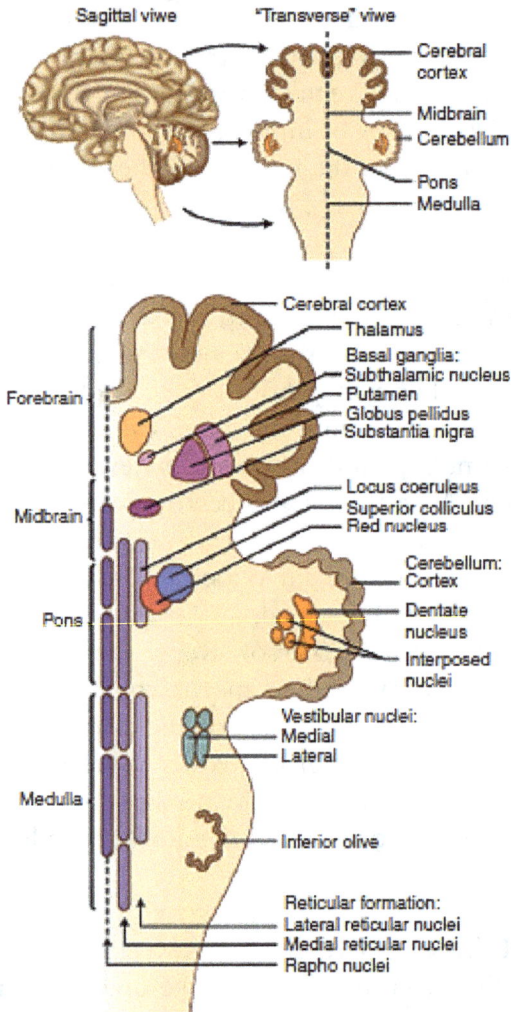

Figure 3.3 From the work of Christopher Miall, University of Birmingham; this figure [adapted from Figure 31.2 of D. Pfaff (ed.), *Neuroscience in the 21st Century*, with permission] offers an idea of the multiplicity and complexity of systems descending from the brainstem for the purpose of motor controls.

voluntary movements. While axons that originated in the cerebral cortex constitute the most direct motor control pathway of this sort, synapsing as it does directly on motoneurons, some pathways that originated in the brainstem perform similar tasks.

An analogy may help one to understand how the cerebral cortex's command to start a movement can be integrated with other controls over body motion. Consider climbing from the top rung of a ladder (the motor cortex) to the bottom (the motor neurons). Motor control pathways in the body can go all the way in one step, can stop at intermediate rungs, and also go down one rung at a time. More recently evolved motor controls, those from the cerebral cortex, are integrated with more ancient motor controls by complex "conversations" on the way down the brainstem and in the spinal cord itself. According to this view, the primary job of the motor cortex is that it integrates other important influences, for example, from the cerebellum and the basal ganglia, so as to make the final, coordinated motor decisions.

Two of the most prominent functions of systems descending from the brain to the spinal cord provide smoothening and coordination of movements, and the ability to generate planned sequences of movements. First, neurons in the cerebellum allow for postural balance and for massive amounts of feedback from our senses and from our muscles themselves, so as to turn potentially jerky, uncoordinated movements into smooth, coordinated movements. In the words of University of Chicago professor Peggy Mason, cerebellar neurons "are perfectly suited to matching outputs to either static or changing conditions and thereby *preventing errors before they happen*" (italics mine). Second, while spinal circuits are clearly capable of carrying out reflex movements, all behaviors that require any thought or planning require the supervision provided by the motor areas of our cerebral cortex and from other areas in our forebrain.

Just in front of the brainstem lie subcortical motor control zones known as the basal ganglia. Compared to the cerebral cortex, the basal ganglia have been heavily conserved through vertebrate phylogeny — that is, among all animals with backbones, from fish to philosophers. I mention these for two reasons. First, failure of dopamine inputs to the basal ganglia causes Parkinson's disease. Second, the bewildering complexity of the basal ganglia — note that they are not yet entirely figured out — is a feature of a control principle in the human brain that you may not have encountered. That is to say, there are two ways of making a nerve cell fire its action potential (its "spike"). Though an excitatory input will do it, if you can inhibit an inhibitory input — so-called "disinhibition" — you also can cause that nerve cell to fire. The principle of disinhibition comes to the fore in the basal ganglia.

Jonathan Mink of the University of Rochester Medical School emphasizes the large number of recurrent loops of motor control in the forebrain, encompassing the basal ganglia, projecting to the thalamus, which in turn projects to the cerebral cortex, which in turn projects back to the basal ganglia. Indeed, neurons in the basal ganglia can markedly increase their firing rates in association with movement, but a lot of their temporal patterns of activity are more complex than that. That is to say, in some cases, basal ganglia neurons are associated with movements more by "releasing pressure on a brake" than by "stepping on the accelerator." Damage to the basal ganglia, as expected, cannot only cause hesitant voluntary movements and unintended involuntary movements but also abnormal eye movements.

In summary, given that mechanisms in the spinal cord actually produce the excitation for movements, we can see that signals from the brain coordinate, regulate, smoothen, strengthen, and program those movements. It will be interesting to see whether advanced robots use the same strategy.

Cerebral Cortex is Required for Voluntary Motion

In the portions of the cerebral cortex dedicated to the control of movement, the firing rates of individual large neurons, and especially groups of neurons, not only predict that a given arm, for example, will move but also the exact direction of movement. The neuron's greatest rate of activity will precede an arm movement in a given direction, and will be absolutely zero before and during an arm movement in the opposite direction. This would be an example of a nonrepetitive, voluntary movement.

All of the brainstem pathways mentioned above must cooperate with behavior regulation by the cerebral cortex. Featuring three different regions of neurons involved in motor control, the cortex's roles stand out as having two special properties. First, the motor cortex provides the finest granularity of muscular control, such as the fine movements of our fingers. Second, as summarized by Miall, cortical neurons are involved in planning complex movements. The easiest way to think about this subject is to assume that cortical controls over movement evolved after brainstem controls, and that the former are effectively superimposed on the latter.

Voluntary movements guided by the cerebral cortex may be regarded as the most complicated of all movements to explain. This is because some scientists feel that motor control systems in the brain contain a kind of internal map — like a three-dimensional Google map — of all the movements it could make, and, in fact, that it represents the movement to itself (in the form of a parallel recurrent pathway) just thousandths of a second before the actual movement occurs. Many neurons whose firing will make muscles contract have axonal branches that inform the relevant sensory systems as to what is about to happen. These are movements that result from an "intention" to act.

In that part of the cerebral cortex primarily devoted to movement, different zones of neurons regulate movements in different parts of the body. Unusually large zones of neurons regulate muscle groups that produce the most finely graded movements, and the reverse is also true. Damaging these cerebral cortex neurons leads to loss of the ability to make precise movements whose force and timing must be controlled within narrow limits. Once the command for any movement is sent out from the cerebral cortex, signaling from any given forebrain motor control neuron is regulated by feedback control from the various senses (for example, pressure on the skin) and, indeed, from the relevant muscles themselves in order to guide that movement to a successful conclusion.

It is exciting to think about how the initiation of movement plays into neuronal circuitry required for human consciousness, which, as defined by neurologists, is the ability not only to be aware of one's environment but also to make purposeful, voluntary movements and to communicate with other humans. Nicholas Schiff, Professor of Neurology at the Weill Cornell Medical School, has based his theory

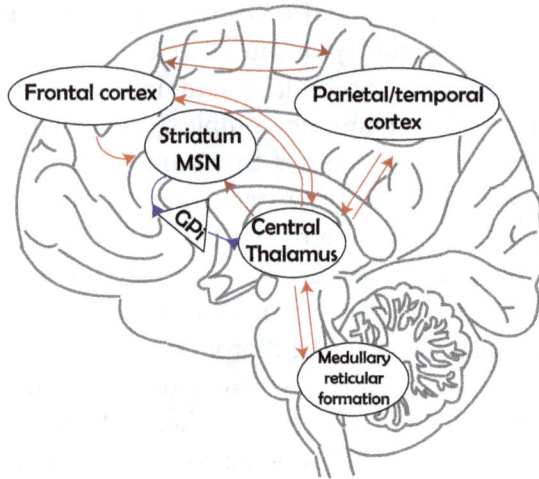

Figure 3.4   A sketch derived from the mesocircuit theory of Nicholas Schiff, M.D., a theory that accounts for the maintenance of consciousness in the normal human brain (references in text). Thalamocortical projections activate the cortex and permit further cortical–cortical cross-excitation. A return circuit back through the basal ganglia supports the mesocircuit by disinhibition.

of a "mesocircuit" (Figure 3.4) on years of studying patients in a dis-ordered condition of consciousness known as a vegetative state, and in particular on the study of high-end vegetative state patients who sporadically show the ability to communicate or follow commands. Schiff's anterior forebrain mesocircuit highlights "key cortical and subcortical components that are vulnerable to effects of severe brain injuries: and widespread loss of normal cerebral connectivity." These are "minimally conscious state" patients.

Schiff has reviewed evidence that traumatic brain injuries cause functional downregulation — a decrease — of activity in mesocircuit structures and that the neurons in those structures even have lowered metabolic activity. His knowledge of (i) the nature of these "patients' brain pathologies," coupled with his knowledge of (ii) human neu-roanatomy and neurophysiology, and (iii) their responses to vari-ous kinds of therapies, led to the mesocircuit hypothesis. Schiff's mesocircuit idea is exciting, because it represents a "superset" — a meaningfully combined and internally communicating group — of all the elements necessary for consciousness and for the initiation of movement.

Human Brain Mechanisms for Repetitive Movements

Tremendous numbers of our movements are stereotyped and repetitive: breathing, walking, all musical activities (because of their constant rhythmic movements), and so forth. Referring to distinctions earlier in this chapter, they can be voluntary (like walking), or involuntary (like obsessive movements in obsessive–compulsive disorder). What neuroscientists have discovered about the neural control of repetitive movements could very likely be applied to robots.

Swedish neurophysiologist Stefan Grillner, Professor at the Karolinska Institutet in Stockholm, has used a variety of electrical recording and behavioral recording techniques to investigate brain mechanisms controlling such movements. His concept, which is most likely to apply to all animals with backbones (all vertebrates, including humans), is called the "central pattern generator" (CPG). This type of mechanism is not limited to one part of the central nervous system. For example, CPGs can appear in the spinal cord, the hindbrain, the midbrain, and the forebrain.

A CPG features both special nerve cells and specific arrangements of connections between those nerve cells. That is to say, in a CPG, the neuroscientist will use electrical recording techniques and see specialized properties of the cell membrane of each individual neuron. In turn, he or she will be able to document complex circular patterns of connections between neurons that may include inhibitory as well as excitatory influences. For humans, CPGs govern the production of a wide variety of repetitive, stereotyped movements with a wide range of time constants, but there is a tendency, true across human populations, for the duration of each episodic movement to last about three seconds.

Grillner's model of how repetitive locomotion occurs highlights the following components that might well apply to behavior controls in robots. First, a "behavior selection" system in the locomotor control areas deep in the forebrain, areas that receive a wide array of inputs, for example, from the cerebral cortex as well as the central thalamus. Second, a "command" system that disinhibits lower motor control neurons, as noted above. In Grillner's concept, the command system can initiate "locomotion by activating the pattern-generating circuits in the spinal cord. Two command systems for locomotion, the mesencephalic (MLR) and the diencephalic (DLR) have been defined. They act via a symmetric activation of reticulospinal neurons which turn on electrical discharges in the spinal circuits."

Then, CPGs themselves are located in the lower brainstem and spinal cord. In Grillner's words, "CPGs contain the necessary timing information to activate the different motoneurons (MNs) in the appropriate sequence to produce the propulsive movements." As I see it, all of these CPGs, in brains ranging from the fish's brain to the philosopher's brain, contain two essential elements. First, there has to be enough neuronal excitation to get the movement started. Second, there have to be inhibitory limits on the CPG's excitation (a) to shift from the left side of the body to the right and vice versa and (b) to adjust CPG activity, as Grillner says, by a "sensory control system, sensing the locomotor movements... to compensate for external perturbations by a feedback action" onto the CPG.

Grillner's approach has tremendous explanatory power for humans and could be applied to robots. On one hand, it is broad. It can be applied to a wide range of repetitive movements among vertebrate behaviors. On the other hand, it is detailed. Grillner's model is rooted in our knowledge of sodium and potassium ion channels through the membranes of individual neurons. Moreover, even several years ago, he himself foresaw how this kind of systematic understanding of repetitive movements of animals and humans could be applied to the engineering of movement control systems in robots.

To control all of these movements, motor control neurons send out a "precommand" signal known as corollary discharge. This signal tells the rest of the nervous system, especially sensory systems, what to expect as a result of the movement that will happen in a few thousandths of a second. As a result of corollary discharge, the person maintains his or her orientation in space and, as well, can make a series of movements in a more controlled, smooth, and well-coordinated fashion.

Routine as our daily repetitive movements are — breathing, walking, etc. — certain ones can be carried too far into obsessiveness. Most familiar may be the psychological need to repeat a common act such as going up stairs one by one, or the repetitive movements of hand-washing. This is called "obsessive–compulsive disorder" by psychiatrists, and is treated with a combination of psychopharmacology (such as antidepressant medication) and cognitive behavioral therapy. More obscure and "exotic" are the repetitive movements of children with autism, a set of symptoms that places these young patients under the range of problems which pediatric neurologists call "autism spectrum disorders." The children may display one of the most distinctive types of repetitive movements called "flapping." The child's arms and hands are extended and move up and down excitedly and rapidly. While it looks as though the young patient may be exhibiting some form of anxiety, the causes of and cures for these repetitive movements await new discoveries in neurological and psychiatric research.

Repetitive movements, by their very nature, are relatively easy for the neuroscientist to study and apply to robots. Much more challenging are the explanations of how we learn new movements, new habits. These would be most exciting to apply to robots because they would provide a new level of flexibility and adaptability for robots in the workplace.

Formation of New Motor Habits

On the other hand, our most interesting movements do not tend to be stereotyped. Reaching for an expensive bottle of wine, scoring a winning basket, or moving closer to kiss a new lover are not things that we do everyday. All of these movements would be under conscious control. So as to be performed in a well-organized and successful way, they must be conscious. All conscious movements require a healthy motor cortex and premotor cortex. They also involve a region of the forebrain underneath the cortex called the "striatum" — because of its striped appearance when viewed through the microscope, looking at brain tissue sections (it is part of the basal ganglia, mentioned above).

Ann Graybiel, Institute Professor at MIT and winner of the National Medal of Science, has delved into the biology of neurons in the basal ganglia and used her findings to construct fascinating theories of how we form new motor habits. These forebrain motor regions called "basal ganglia" because of their position in the human brain, do not send their axons to motor neurons or even to the spinal cord. Instead, they communicate powerfully with the cerebral cortex and with upper levels of the brainstem. They are complex, in that they feature loops within loops of circuitry. Not just neuronal excitation but also disinhibition is the name of the game — turning off inhibitory inputs can have results similar to excitation! Instead of directly commanding movements, they have a "permissive role" in the initiation of movements.

Graybiel runs one of the world's leading laboratories in the investigation of how we learn, maintain, and change behavior. She studies the striatum and the cortex, and talks about learning behavioral "habits." Paraphrasing neurophysiologist Sten Grillner, Graybiel has almost single-handedly demonstrated a modular organization within the striatum. Some groups of neurons known as striosomes receive inputs from older parts of the cerebral cortex and are linked up with dopamine neurons.

Graybiel argues that once they are learned, "deliberate behaviors become routine." Electrical recordings of nerve cells in the striatum during the formation of a habit — rats running a maze in her MIT lab — gave surprising results. At the beginning, these neurons in one part of the striatum, the "habit striatum," "fired" their electrical signals all the time during the maze run. When the behavior was well learned, however, these neurons were active just at the beginning and the end of the run. But, in a different part of the striatum, exactly the opposite dynamics were observed. In this other "deliberation" part of the striatum, as the animals were in the middle of learning their task, these "deliberation" neurons were most active in the middle part of the maze run, especially around the time when the animals had to make a behavioral choice. To paraphrase Graybiel, when the learning process is finished, forebrain circuits began to treat the maze-running habit as a single "chunk," or unit of behavior.

Of course, the laboratory animals were running the maze in order to get a reward. How does reward-seeking work? Modern neuroscience has proven that the small molecule constituting the neurochemical dopamine plays a central role in rewarding the performance of newly learned tasks. What is crucial is the production of dopamine by neurons in the midbrain and its arrival, following the axons of those neurons, in the forebrain. Graybiel and her team expected sudden spikes in dopamine upon presentation of the reward. Instead, quoting her, "to our surprise, instead of mainly finding isolated dopamine transients at the initial cue or at goal-reaching, we primarily found gradual increases in the dopamine signals that began at the onset of the trial and ended after goal-reaching." Thus, under these circumstances, rather than dopamine release suddenly and instantly stamping in a reward, prolonged dopamine signaling could provide sustained motivational drive, "a control mechanism that may be important for normal behavior and that can be impaired in a range of neurologic and neuropsychiatric disorders."

In turn, certain neurons in the striatum keep track of whether a particular behavior has been successful or not in obtaining an award. These are a particular subgroup of neurons that have a steady background level of firing rate. Thus, they are called "tonically active neurons." Some of these neurons — probably cells that use the neurotransmitter acetylcholine — fired intensely when the animal had reached the goal of the learning task and was receiving a reward.

Graybiel and her team consider electrical activity in the striatum, this important subdivision of the basal ganglia, to be a starting point rather than a finishing line. That is to say, once a "chunk" of habit is learned, it can be altered or linked to other behaviors. What happens when a habit has to be modified? In Graybiel's concept, the cerebral cortex is monitoring the performance of the striatum. "Tweaking" the activity of the neocortex can interrupt a habit, thus opening the way for breaking bad habits and learning new habits.

It is very unlikely that circuitry for robotic learning will be designed through attempts to mimic the sophisticated mechanisms discovered by Graybiel's team. The circular arrangements of some of the basal ganglia circuitry and the delicacy of the "tweaking" by the cerebral cortex will impress engineers as designs that are too likely to mess up. Robot engineers will have to do something else.

So, Ann Graybiel's work has told us a lot about how new habits are formed. And, as I have briefly sketched, we know a lot about how movement is produced by the human brain, a lot about how repetitive movements are produced by CPGs, and about habit formation. But suppose the environment changes and the learned habit produces unfortunate results, that is, errors?

Reza Shadmehr's group at the Laboratory for Computational Motor Control in the Biomedical Engineering Department at Johns Hopkins asked the following question: How does the brain alter behavior after experiencing an error? Their answer surprised everybody. It turned out that their results revealed a type of memory not recognized before — that of the motor errors themselves. I address this because Shadmehr's basic concept could apply, one for one, to memory for motor errors by robots.

In Shadmehr's studies, three groups of subjects were instructed to perform a task in which they were reaching for a target while their arm movements were perturbed by a force that changed at a slow rate or a fast rate. Slow rates of change permitted subjects to learn more from their errors than in fast-switching environments. For some subjects the rate of force alteration changed across the experiment from slow to fast; for other subjects, it was the reverse. In the former group, error sensitivity decreased, while in the latter group the reverse occurred. A critical step in their work derived from their prediction that if the brain really carries a memory specifically of errors, "it should be possible to simultaneously increase sensitivity for one error while decreasing it for another." Indeed, in a complex experimental design their results showed that.

Shadmehr concluded that in addition to a motor memory for an act and a memory for perturbations, the human brain uses a memory specifically for movement errors.

Thus, neurobiologists now know not only about how habits are learned, but also about how errors are remembered. As mentioned during my summary of motor controls above, most of our own motor control mechanisms are likely beyond robot engineering, although some of the basic concepts might help with robot motor control.

I will now highlight what I consider to be two obvious and irreconcilable differences between human's motor circuitry and robots'. Working around those differences, robot designers of the future will have to answer two basic questions. First, they will have to decide what aspects of human circuitry they want to reproduce (or mimic) in robots. Second, and even more important, they will try to compensate for undeniable shortcomings in human performance; in doing so they will create human–robot synergies.

Differences Between Humans and Robots

The complexity of motor controls in the human brain tells us immediately that it is unlikely that robots will have similar control systems. Both types of systems can get many jobs done, but not in the same way.

Here is a crucial point: precisely because the "guts" of motor control systems will be different between humans and robots, the respective strong and weak points will differ considerably. Planning for any given task, roboticists can look at the weakest aspects of the humans' contributions to achieving the task and use the fact that their robots do not have corresponding weaknesses, thus contributing to a stronger overall performance. Therefore, as noted, we can anticipate synergies between them.

In addition to the differences in motor control circuitry, robots will not for the foreseeable future have any elements like the marvelous mammalian nerve cell. Wondrous not only in its chemical and structural complexity but also in its adaptability, the typical neuron hides some of the secrets of how the human brain preforms its tasks. Figure 3.5, cartooning a neuron, just begins to hint at the marvellousness of the single nerve cell, variegated, complex, and harboring the possibility of useful modification.

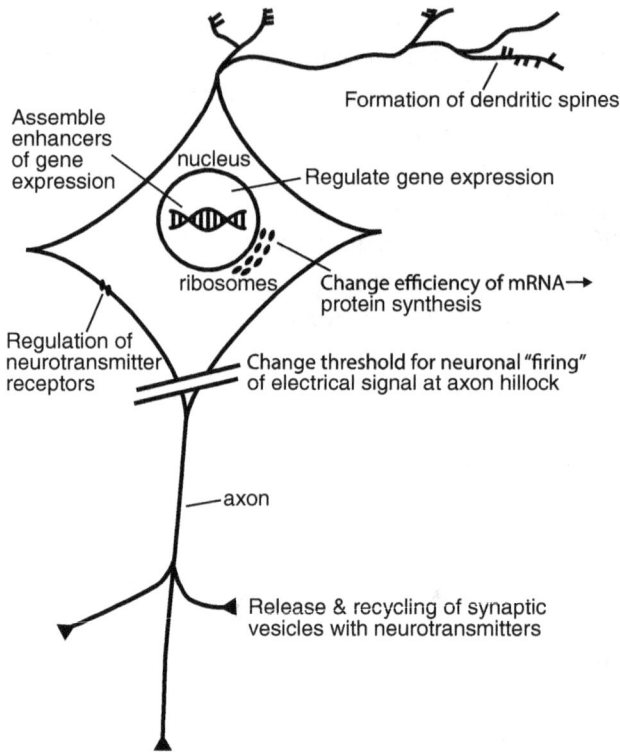

Figure 3.5 The marvelous complexities of neurons derive not only from their precise morphologies and their patterns of electrical signaling, but also from their large number of internally regulated properties. Just a few are sketched here.

One of 2014's heralded accomplishments was the development of a "brain-like chip" at IBM. The main idea is that the so-called "neuronal mimics" known as TrueNorth can signal to each other when their inputs have reached certain thresholds. The problem, as admitted by the chief of the program, is the cost of power for the overall computer. Compared to our "amazing neurons," TrueNorth uses too much power. As IBM makes them faster and faster, the power shortfall becomes worse. That is to say, the fastest chips use the most power. The human brain's performance is not constrained in this way.

Table 3.1 demonstrates that we should hardly be surprised by human–robot differences in motor controls.

Table 3.1   Comparison of Human and Robot Motor Controls*

	Human	Robots
Evolution	Biological	Industrial design
Constraints	Material strength, speed	Economic, marketplace

*Parallel to Chapter 2.

Evolution

The renowned geneticist Theodosius Dobzhansky, the late professor of genetics at Rockefeller University, observed that "nothing in biology makes sense except in the light of evolution." Of course, he meant *biological* evolution, in which modifications of an organism's DNA usually harm that organism's reproductive fitness but occasionally increase it and end up in the germ line — that is, in the sperm or eggs so that it will be passed onto the next generation. By this accidental process, gene expression and subsequent protein expression in the cells of that organism become better suited to that particular environment.

Evolution of robot motor controls obviously cannot follow the same process as biological organisms. Robots do not have DNA, and any proteins (such as fibrous proteins that might be used for a combination of flexibility and strength) that show up in their various parts were introduced purposefully by industrial chemists. Research and development groups in companies that produce robots have put in place processes in which design plans "evolve" — processes that are no less complex than the human mind and, for that matter, groups of human minds. If there are any "rules" to describe robot design, they have nothing to do with biological evolutionary mechanisms.

Human Constraints are not Robotic Constraints and Vice Versa

Likewise, consider the constraints. Any human athlete that examines his or her performance limitations will struggle to produce movements that exhibit more strength, speed, and coordination. For most of us, constraints on our production of motion imposed by neural and muscular materials are all too obvious.

However, robot metallic limbs moved by robot motors and modern electronic controls can have ultimate strength, lightning-fast movements, and computer-coordinated movements. The limitations — differing from those on human movements — are mainly financial, reflecting the adequacy of the product for market demands. Some would say that, given adequate financing, robotic movement controls will almost always achieve greater precision than their human counterparts. Others would raise the distinction that neuroscientist Daniel Wolpert makes: that while computers can match or beat human thinking (such as IBM's Watson playing chess), even young children easily outperform the world's best robots in terms of smooth manipulation of chess game pieces. Motor control is really difficult.

Everybody has problems, and that sentiment certainly applies to the behaviors of humans and robots. In fact, comparing the problems to be solved in human and robot motor control offers us one of the interesting ways of comparing the two. As you might expect, this comparison has piqued the interest of engineers working for profit-making companies. Dan Washkewicz, the chief executive officer of a highly experienced group Parker Hannifin, now wants to make "motoric robotic braces that support, bend and move the legs of people who can't walk on their own because of spinal cord injuries, multiple sclerosis or strokes," according to Bob Tita, writing for *The Wall Street Journal*. According to Tita, an Israeli startup, ReWalk Robotics, already has FDA approval for similar efforts. On one hand, this work promises to spawn a multimillion-dollar industry. On the other, it has already and will continue to call for the skills of the best academic engineers, such as Michael Goldfarb of Vanderbilt University. Patients' personal stories are compelling. One Georgian (quoted by Tita), whose spinal cord was damaged by a motorcycle accident, not only can walk but has lost his painful muscle spasms and strengthened his core muscles by using a wearable robotic device. This will be big news — medically, commercially, and ethically.

For the human brain, the chemistry is tricky and the spatial precision required is on the order of thousandths of a millimeter. Further, it will always be a question as to whether our neuromuscular systems can produce the strength and speed of response adequate for the task. On the other hand, with robots, strength and speed are no problem. In seeking to meld the two, can we get the best building materials? And what about economic costs?

Machine-Assisted Human Movements; Direct from Brain to Prosthesis

Many current projects are using robot engineering and artificial intelligence to extend human capacities, especially for patients whose bodies have been badly damaged by war, disease, or accidents. In spite of the contrasts between human movement controls and robot motor controls, for example, as in Table 3.1, robot engineers clearly will get robots to do many things that humans can do, and more. Consider one special application of automated motor control. Geoffrey Ling, a medical doctor in charge of a large program at the Defense Advanced Research Project Agency (DARPA), has the task of designing man–machine systems by which injured military personnel can control artificial limbs. He has succeeded.

DARPA is concerned with restoring motor function in military personnel who have returned from battle with injuries of the central nervous system. Under Ling's leadership, it has facilitated the development of computerized systems that allow patients to use brain signals to control artificial limbs.

Miguel A.L. Nicolelis runs one of the leading brain/machine laboratories at Duke University and in Sao Paolo, Brazil. He has achieved eye-popping results, such as when his experimental monkeys produce neural signals that are transmitted to control movements by a robot in Japan. He likely acquired the courage to begin neural/prosthetic engineering by becoming an expert in neural population coding. That is to say, a scientist who already knows a lot about the signaling dynamics in the primate cortex has a sense of "what is going on there," and as a consequence can try to represent and employ that signaling in a useful, even therapeutic bioengineering context. Nicolelis' lab recorded specific patterns of electrical activity from the surface of the cerebral cortex as the subjects carried out a throwing motion. Duplicating those specific patterns enables a device controlling an artificial arm to throw a ball — in effect, throwing by just thinking about it.

The lab has now invented an exoskeleton whose physical strength could make up for the bodily weakness of a paraplegic. Controlling the movements of such a structure could allow paraplegics to leave behind their wheelchairs and walk, at least under certain circumstances. This has worked well at the University of Houston lab of Jose Contreras-Vidal, but the machinery has weighed as much as 80 pounds. However, exoskeletons can now be made so soft and lightweight that Conor Walsh, the chief of one DARPA project, can have his subjects wear them under regular clothes. Injuries to humans can be compensated for by the types of engineering used in robots.

The work goes well beyond activating motor neurons. Choosing where to do your work in the body is the first step. Should it really be in the brain, or might controls going directly to the peripheral nervous system be better? This work involves materials scientists as well as nerve cell growth experts. Experts are using transplanted nerve fibers and getting them to grow long enough to serve the goal of "covering the gap between the host's axons and the electronics." If the electronics were flexible and nontoxic to surrounding tissue, the neuroscientists' job would be easier.

Ken Shenoy, in the Department of Electrical Engineering at Stanford, takes a similar view. His "perspective concentrates on describing the neural state using as few degrees of freedom as possible and on inferring the rules that govern the motion of that neural state."

Based on this kind of research, an entire field has developed — called "noninvasive orthopedics," the development of exterior skeletal structures and artificial limbs that could help patients with a wide variety of serious injuries. But that field just addresses the easy questions. The profound questions deal with how seriously and deeply we can understand the brain signals that control normal body movement, such that patients' control over various prosthetic devices moves beyond the primitive.

In the words of Bensmaia and Miller, regarding spinal cord injury, "brain/machine interfaces might enable a patient to exert voluntary control over a prosthetic or robotic limb or over the electrically induced contractions of paralyzed muscles. A parallel interface could convey sensory information about the consequences of these movements back to the patient." A pioneer in this area was John Donoghue, Director of the Institute for Brain Science at Brown University. His extensive laboratory develops "Motor Neural Interface Systems (NIS), which aim to convert neural signals into motor prosthetic or assistive device control, allowing people with paralysis to regain movement or control over their immediate environment."

As a young boy, Donoghue suffered from the same disease as my son: Legg–Perthes disease, which prevents a child from walking for two or three years. Years ago he used intraoperative recordings of the human cerebral cortex to deduce what kinds of signals would lead to the planning and production of coordinated movements. His first striking results treated a tetraplegic human in whom neural activity "recorded through a 96-microelectrode array implanted in primary motor cortex demonstrated that intended hand motion modulates cortical spiking patterns three years after spinal cord injury. Decoders were created, providing a 'neural cursor' with which the patient could open simulated e-mail and operate devices such as a television." This achievement, and the work cited below, offer examples of how combinations of electrophysiology and the kind of artificial intelligence circuitry used in robots can improve human lives, especially those of people who have suffered severe damage through war or accidents.

At the University of Pittsburgh laboratory of Andrew Schwartz, artificial limbs can be controlled by the patient's thoughts. The laboratory of Michael McLoughlin in the Applied Physics Lab (APL) at Johns Hopkins has likewise focused on upper extremities as an especial challenge because of the wide variety of activities normally carried out by hands and arms. In the future, from scientists like Donoghue and McLoughlin, you can look for wireless systems in which brain signals will be transmitted to devices worn on the patient's body, devices that control movements of all four limbs. The robots themselves, produced by companies like Boston Dynamics, will walk and run "like living creatures." A researcher at that company, Marc Raibert, uses the concept of dynamic balance to control continuous motion that allows his robots to remain upright. His robots are not yet ready for the mass market, but they exemplify what is coming.

The APL works with collaborators around the country. Collaborating with neurobiology professor Andrew Schwartz, a neurologist at the University of Pittsburgh, they implanted two 96-channel intracortical microelectrodes in the motor cortex of a tetraplegic patient. Brain/machine interface training was done for 13 weeks. After that, the patient was able to move the prosthetic limb freely in the three-dimensional workspace on the second day of training. Further, collaborating with scientists at the University of Chicago, the APL team explored the issue that "tactile sensation is critical for effective object manipulation, but current prosthetic upper limbs make no provision for delivering somesthetic feedback to the user." To remedy this shortcoming they developed a prosthesis whose finger contained a pressure transducer where the force output is converted into a regime of intracortical microstimulation through long-term implanted multi-electrode arrays. Activation of those electrodes "stands in" for natural somatosensory inputs. Moreover, the team has gone on to show that the performance of a tactile task by experimentally prepared monkeys is equivalent whether stimuli are delivered to the natural or the prosthetic finger.

Cortical controls over straightforward movements of the limbs may be more complicated than you would imagine. Kenneth Shenoy, a bioengineer at Stanford University, in his *Annual Review of Neurosciences* (2013) article, talks not just about the firing of a cortical neuron causing a motor neuron in the spinal cord to fire and the limb to move, but also about what he calls a "dynamical systems perspective" in which "activity in the motor system reflects a mix of signals: some will be outputs to drive the spinal cord and muscles, but many will be internal processes that help to compose the outputs but are themselves only poorly described in terms of the movement. They may, for instance, reflect a much larger basis set of patterns from which the eventual commands are built." To give you an idea of how complex this could get, consider his definition of a population dynamical state: "a set of coordinates, often represented as a vector, describing the instantaneous configuration of a dynamical system and that is sufficient to determine the future evolution of that system and its response to inputs. The population dynamical state of a neuronal network might be the vector of instantaneous firing of all its cells or may incorporate aspects of the neurons' biophysical states. It may also be a lower-dimensional projection of this network-wide description." In view of this modern, still evolving approach to motor control systems in the human brain, it is likely that matching the sophistication of robot movement controls to human motor physiology will become ever more daunting.

This field is developing rapidly and is not limited to the US. I have already mentioned labs in Brazil and Japan, and when I lectured at the annual meeting of the Turkish Neuroscience Society I found that Emec Ercelik at the Istanbul Technical University had used a mathematical model based on basal ganglia circuitry to produce a robot that could find food, recognize it as food, and lift it and place it into its "mouth."

Based on all of this work, it is possible to envision paths toward ever more success both for robots themselves and for how motor control engineering can help injured humans. The likelihood of this type of human–robot interaction makes it even more important to discuss an area of robot engineering and human neuroscience seemingly more difficult to think about: emotions. The next chapter will discuss our growing knowledge of how emotions are regulated by the human brain.

Summarizing the elements of a rapidly developing field of neuroscience, the human brain uses both local connections and long distance axonal projections not only to produce movement but also to plan and to smoothen well-coordinated movements. Robot motor controls differ markedly from human ones both in their evolution and in their constraints. The intricacy of human motor controls differs markedly from emotional regulation, as the next chapter will explain.

Further Reading

Atallah H *et al.* (2014) Neurons in the ventral striatum exhibit cell-type-specific representations of learning. *Neuron* **82**: 1145–1157.

Bensmaia S, Miller L (2014) Restoring sensorimotor function through intracortical interfaces. *Nat Rev Neurosci* **15**: 213–235.

Burguière E, Monteiro P, Mallet L, Feng G, Graybiel AM (2014) Striatal circuits, habits, and implications for obsessive–compulsive disorder. *Curr Opin Neurobiol* **30C**: 59–65.

Burke R (2012) Spinal motoneurons. In: Pfaff D (ed.), *Neuroscience in the 21st Century*. Springer, Heidelberg, New York.

Butt SJ, Harris-Warrick RM, Kiehn O (2002) Firing properties of identified interneuron populations in the mammalian hindlimb central pattern generator. *J Neurosci* **22**: 9961–9971.

Fridman EA, Schiff ND (2014) Neuromodulation of the conscious state following severe brain injuries. *Curr Opin Neurobiol* **29**: 172–177.

Graybiel A, Smith K (2013) Good habits, bad habits. *Sci Am* **310**: 38–43.

Grillner S, Markram H, De Schutter E, Silberberg G, LeBeau FE (2005) Microcircuits in action — from CPGs to neocortex. *Trends Neurosci* **28**(10): 525–533.

Grillner S, Kozlov A, Dario P, Stefanini C, Menciassi A, Lansner A, Hellgren Kotaleski J (2007) Modeling a vertebrate motor system: pattern generation, steering and control of body orientation. *Prog Brain Res* **165**: 221–234.

Herzfeld D *et al.* (2014) A memory of errors in sensorimotor learning. *Science* **345**: 1349–1355.

Howe M, Tierney P, Graybiel A (2013) Prolonged dopamine signaling in striatum signals proximity and value of distant rewards. *Nature* **500**: 575–579.

Howe MW, Atallah HE, McCool A, Gibson DJ, Graybiel AM (2011). Habit learning is associated with major shifts in frequencies of oscillatory activity and synchronized spike firing in striatum. *Proc Natl Acad Sci USA* **108**(40): 16801–16806.

Kandel ER *et al.* (eds.) (2013) *Principles of Neural Science*, 5th edn. McGraw-Hill, New York.

Kiehn O, Kjaerulff O (1998) Distribution of central pattern generators for rhythmic motor outputs in the spinal cord of limbed vertebrates. *Ann NY Acad Sci* **860**: 110–129.

Kozlov A, Huss M, Lansner A, Kotaleski JH, Grillner S (2009) Simple cellular and network control principles govern complex patterns of motor behavior. *Proc Natl Acad Sci USA* **106**(47): 20027–20032.

Lemon R (2008) Descending pathways in motor control. *Ann Rev Neurosci* **31**: 195–218.

Mason P (2011) *Medical Neurobiology*. Oxford University Press, Oxford.

Nicholls J *et al.* (eds.) (2001) *From Neuron to Brain*, 4th edn. Sinauer, Sunderland, MA.

Pfaff D (editor). (2012) *Neuroscience in the 21st Century*. Springer, Heidelberg.

Purves D *et al.* (eds.)(2012) Neuroscience, 5th edn. Sinauer, Sunderland, MA.

Schiff ND (2010) Recovery of consciousness after brain injury: a mesocircuit hypothesis. *Trends Neurosci* **33(1)**: 1–9.

Smith K, Graybiel A (2014) Investigating habits: strategies, technologies and models. *Front Behav Neurosci* **8**: 39–52.

Smith KS, Graybiel A (2013) A dual operator view of habitual behavior reflecting cortical and striatal dynamics. *Neuron* **79(2)**: 361–374.

Squire L *et al.* (eds.) (2013) *Fundamental Neuroscience*, 4th edn. Academic Press/Elsevier, San Diego.

CHAPTER 4

Emotions

This chapter will discuss our knowledge of how the brain is "wired" such that we experience and express emotions. Surprisingly, brain mechanisms for emotion are simpler, grosser, and less articulated than the sensory and motor mechanisms described in Chapters 2 and 3. A sad token of this fact is that I have seen brain-damaged patients who cannot move or talk or respond to most stimuli but can respond to a stimulus with emotional content. Emotions are primitive. So it is likely that matching robots' "emotion-like" behaviors to what humans expect of emotions may actually be easier than matching sensory and motor mechanisms. Just to emphasize — I do not argue that robots will "have" emotions, but that their behaviors will *emulate* emotions.

Neuroscientists know a lot, at least about some emotions, and can actually map the mechanisms underlying them. In discussing the brain's emotional wiring in this context, I wish to suggest that *since* emotions are the consequence of specific brain mechanisms, we can duplicate those mechanisms ("wiring") in robots. To make this point, I will describe what we know about a range of emotions, not because in every case we will want robots to display these emotions but because I want to demonstrate our advanced understanding of why and how emotions occur. What matters in the discussion that follows is not that you draw a one-to-one connection between what a human and a robot might experience but, rather, that you appreciate how skilled we already are in reverse-engineering the emotional components of the human brain. Thus, where I talk about fear and anger, there is a good chance that we will want robots to display comparable reactions (for example, as likely error messages); where I talk about lust, there is less chance that robots will ever display comparable reactions (even though we fully expect robots to reproduce themselves).

Scientists know exactly what is going on — and can specify brain mechanisms in a transparent manner — when a person focuses his or her eyes on a tree or even follows a bird in flight. During the past few decades modern neuroscientists have worked hard to map these operations, to follow them in a detailed, logical sequence of brain mechanisms. We know a lot about what changes in the brain when we become afraid or turn vengeful. However, we do not define emotions as precisely as we do specific sensory stimuli or specific motor behaviors.

A philosopher or psychologist might adopt a commonsense definition of "emotion" such as that offered in Wikipedia: "Emotion is a subjective, conscious experience characterized primarily by psychophysiological expressions, biological reactions, and mental states." A neuroscientist takes less interest in defining the boundaries of what is meant by emotion than in explaining brain mechanisms that produce feelings and behaviors that all would agree are "emotional." The next chapter will argue that programs in robots could produce apparently emotional behaviors; Chapter 6 will point out that understanding such emotion-like behaviors will be important for human–robot interactions.

Emotional controls by the brain differ greatly from sensory responses and the regulation of motor behaviors. In the words of Eric Kandel, a Nobel Prize–winning neurophysiologist, " Gaining biological insight into emotion is a two-step process. First, we need to understand a psychological foundation for analyzing an emotion, and then we need to study the underlying brain mechanisms…"[Kandel (2012)]. For the first step we get help from the Smith and Kosslyn (2007) text and from many academic psychologists. They use the term "emotion" to refer to "mental and physical processes that include aspects of subjective experience, evaluation and appraisal… and bodily responses such as arousal and facial expressions." Of course, in laboratory animals, we must study the objective behaviors that, in humans, would correspond to a particular subjective feeling. So, in the case laboratory studies of fear, an expert like New York University professor Joseph LeDoux would describe himself as analyzing the neuronal basis of "defensive responses to threat."

In the brain, emotional signals report on the state of the organism, the overall condition of the individual. As such, their neuroanatomical pathways differ markedly from the straightforward sensory and motor controls previously discussed. Instead of point-to-point hookups for precise perception and nuanced muscle response, the neuronal groups involved in emotion get inputs converging from a broad range of brain regions, and "emotion neurons" radiate their outputs to many different sites in the brain (Figure 4.1). Whether this or that muscle contracts when a given emotion influences behavior depends on the situation. That is to say, the emotional mechanisms' outputs radiate and can influence many, many muscles in a nonlinear manner. Here is the important distinction: the emotional state will tell the animal or human to either *approach* that stimulus or *avoid* it. The

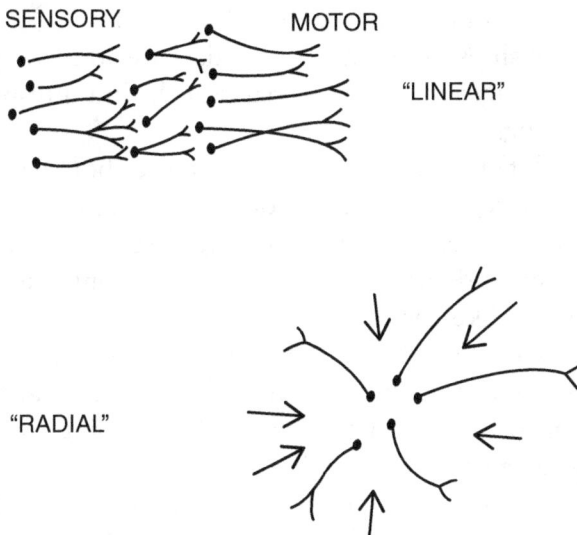

Figure 4.1   While sensory systems (Chapter 2) must include a predominately "linear" signaling structure to get from receptor cells to decision sites in the brain, and motor systems must do similarly (to get signals from motor commands to motor neurons), emotional systems in the brain are different. Here I have sketched them as having a "radial" pattern to indicate that they get many divergent inputs in order to assess the state of the body, and respond with many divergent outputs that modulate a wide variety of behavioral, autonomic, and hormonal processes.

approach/avoid dichotomy summarizes the most fundamental choice in emotionally guided behaviors.

That dichotomy is intriguing, because the most elementary question we can ask about a human–robot interaction is: "Does the robot understand, go along with , become 'happy' with what the human wants it to do?" versus "Does the robot, instead, react against the instructions coming from humans and avoid the situation at hand?"

Emotions used to be off-limits to serious neurophysiologists. The renowned neurologist Antonio Damasio, was told at the beginning of his career to avoid the study of emotion because "there's absolutely nothing there of consequence." Since then he and his wife, Hanna, have demonstrated how the brain attains a feeling for the body to which it is attached. In his words, "mind begins at the level of feeling." So things have changed regarding the analysis of brain mechanisms for emotion. Instead of concentrating purely on what might be called the "Apollonian" brain (named after the Greek god of wisdom, Apollo; parts of the brain responsible for cool, logical stimulus–response relationships), more neuroscientists are addressing the operations of the "Dionysian" brain — parts of the brain responsible for desires and feelings).

We know a lot and are quickly learning more about nerve cells that participate in the regulation of emotions. As a result, we will know how people feel by measuring their biological signals; we can integrate those types of measurements into circuitry that would be available to robots. Engineers like the Stanford University Xbox team develop game consoles with transducers to monitor breathing, skin sweating, heart rate, etc. The Affectiva company of Rosalind Picard and Rana el Kaliouby, in the Media Laboratory at MIT, doing similar things, will quantify at least some emotional states in a manner that will empower social robots. Emotional facial expressions will be read by robots, organismal states will be reported, and emotion-like postures and noises will be emitted by robots.

Mechanisms

Here I sketch some of the mechanisms for emotions connected with objectively measured animal behaviors. I recognize clearly that, as pointed out most prominently by New York University Professor Joseph LeDoux, an animal's behavior measured precisely in the laboratory is not the same as a human feeling. The domain of human feelings allows for levels of consciousness and psychological causation that we do not suppose to be present in the mouse brain. But to let that realization stand in the way of scientific thinking, experimenting, and writing would amount to a purely obscurantist attitude. Some relations between laboratory animal behavior and human emotion will be tighter than other animal–human relations. Below, I concentrate on emotional behavioral systems that most obviously translate from the animal brain to the human central nervous system.

Prominent are neurons in three brain regions: the amygdala, the hypothalamus, and the hindbrain reticular formation. In both the lab animal brain and the human brain, these three regions feature the radiating patterns of connections illustrated in Figure 4.1. All three are hidden from view in typical pictures of the human brain. The amygdala has been rolled underneath more modern cortical tissue during the evolution of the forebrain. It is tucked next to the brainstem as the six-layer neocortex that makes us human became more elaborate. The hypothalamus — and its neighbor, the preoptic area — are just above the roof of the mouth (Figure 4.2). The hindbrain reticular formation is usually rendered invisible in typical pictures of the human brain.

Highly motivated behaviors associated with emotions feature what the Nobel Prize–winning biologist Niko Tinbergen called "action-specific energies" (Figure 4.3). Beneath all those specific forms of emotional energies lies a primitive, deep, and fundamental force for the initiation of emotion-like behaviors. My Harvard book called that "generalized arousal," and the large cells of the hindbrain reticular formation serve that fundamental and essential function *par excellence* (Figure 4.4). I cannot think of any reason that this structure for producing emotion-like behaviors cannot be emulated in robot designs of the future.

Fear

Fear — trembling in place, running away, anticipating trouble — is a huge topic. Brilliant neuroscientists have concentrated on brain

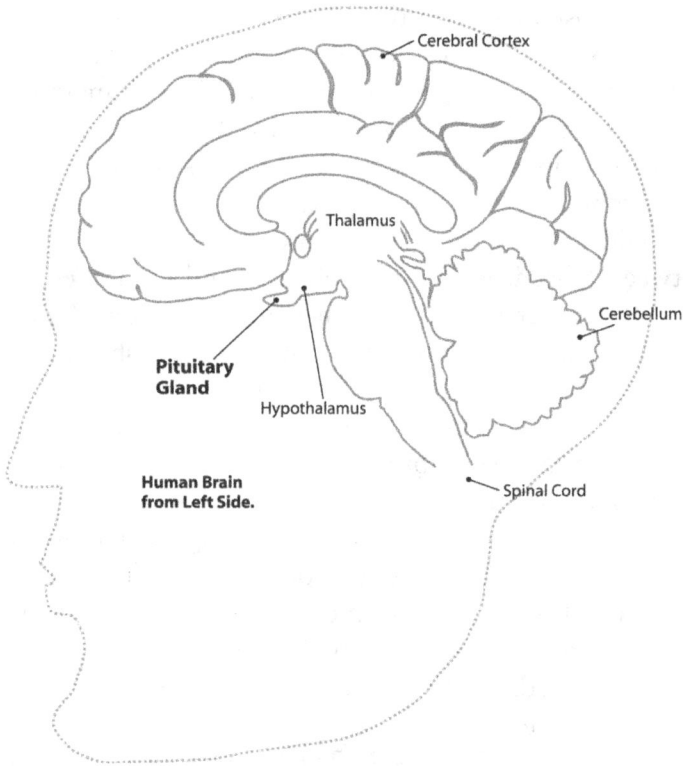

Figure 4.2 Looking at the human brain from the left side, the hypothalamus is indicated, and the preoptic area is just anterior to it (to the left, along the bottom of the brain).

mechanisms that produce fear-like behaviors in laboratory animals. From the perspective of this book it is important to know something about these mechanisms because the depth of neuroscientists' understanding convinces me that skilled engineers could both program fear-like circuitries into robots of the future and, correspondingly, understand where robot behavior is "coming from" when a robot exhibits fear-like behavior.

The feeling of fear depends on the electrical activity of neurons in the human amygdala. Neuroanatomists thought of the amygdala as a single structure, so named because, when viewed in the human brain from below, it is shaped like an almond (the Greek word for "almond"

COGNITIVE FUNCTION   **EMOTIONAL FUNCTION**

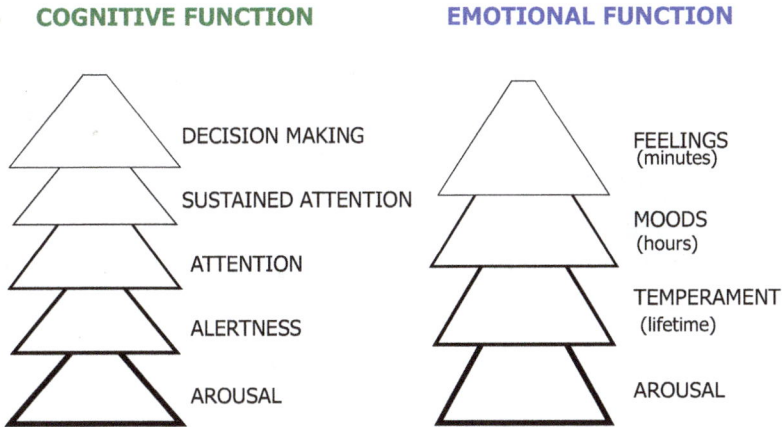

Figure 4.3 Human brains have complex arousal systems that support all cognitive functions and all emotional expressions. For instance, you can be aroused without being alert, but not the reverse. Likewise, the intensity of all emotional expressions, whether through temperament, mood, or momentary feeling, depends on brain arousal systems.

is "*amydale*"). Now we know that the amygdala has more than ten subdivisions with separate roles. Rather than delving into all of those, let us take a straightforward, three-step approach to explaining fear:

(i) How do danger signals get to the amygdala?
(ii) How do fear messages radiate out from the amygdala?
(iii) What are nerve cells in the amygdala doing to execute that transformation from input to output?

First, sensory signals that warn us of danger converge on the amygdala using a variety of routes. We mainly divide these signals according to whether they go through the upper brainstem region known as the thalamus (which is also the entryway for signals headed for the cerebral cortex) or avoid the thalamus. The latter, nonthalamic signals, come both from the classical olfactory system which handles an infinity of different smells, and from the more specialized "accessory olfactory" system which handles pheromones, among other odors. Thalamic neurons give all the other danger signs: these could be, for example, visual, auditory, taste, or touch. In modern neurobiology, the auditory warnings of danger are the most famous because LeDoux

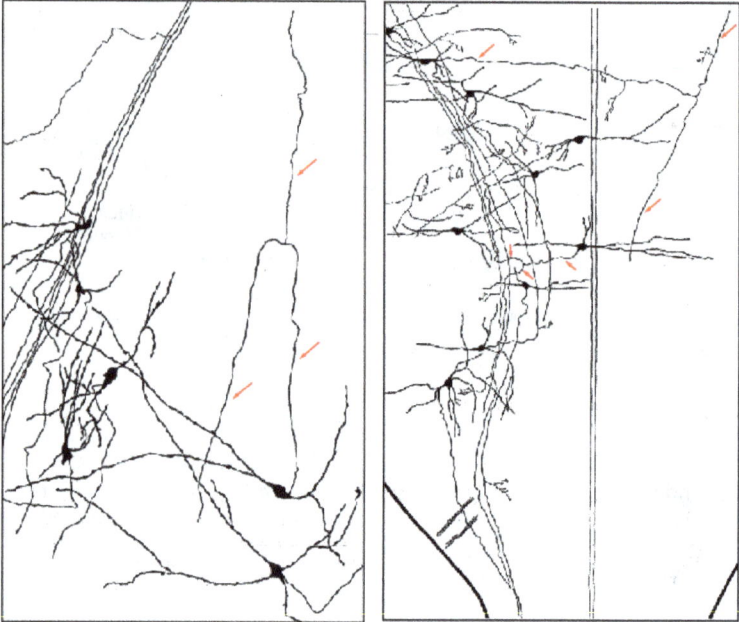

Figure 4.4 Large neurons in the medullary reticular formation. Cell bodies are black. Bifurcating axons that can both control movement and activate the forebrain are marked with arrows. [Figure 2.6 adapted from: Pfaff D, *Brain Arousal*, with permission.]

has been so successful in analyzing exactly how they work. He has shown how a sound that warns the animal of a painful foot shock can use what he calls the "low road," a more ventral route directly from thalamus to amygdala, or the "high road," a more dorsal route which loops up into the cortex on the way to the amygdala. He characterizes the low road as faster and the high road as more accurate. Whichever, these enter the amygdala in its lateral subdivision. Similar multilevel routes to fear could easily be programmed into robot circuitry. The more intricate the robot circuitry for fear, the greater the finesse and subtlety of robot "tear responses."

Second, when the amygdala has done its job, what happens? Amygdaloid messages radiate all over the place, accomplishing a variety of tasks which we associate with fear. Axons going to the hypothalamus can activate our autonomic nervous systems — blood pressure, heart rate, sweating, and so forth — and can influence hormonal outputs from the pituitary gland, which dangles from the bottom of the hypothalamus. The hormonal changes can be fast, because the tiny nine-amino-acid vasopressin can protect us from blood loss if we are wounded. They can also be slow, since the large protein hormone adrenocortical trophic hormone (ACTH) will turn on the adrenal gland, releasing stress hormones. Some amygdaloid axons take a long route all the way down to the midbrain. Neurons found there in the middle of the midbrain — the "central gray" — cause the immediate reflex response to fear. A laboratory animal anticipating painful shock will freeze immediately, sometimes for quite a long time, before taking the next step. Amygdaloid axons going to our cerebral cortex account for the obvious intellectual or psychological components of fear.

Between inputs and outputs of the amygdala, nerve cells with extremely short axons (the output projections from nerve cell bodies) are getting the job done within the amygdala itself. The axonal projection from the lateral region of the amygdala into the central region (the "central nucleus") is the best established. David J. Anderson, Professor of Biology at Caltech, has used the most modern genetic and molecular techniques to piece together how neurons within the amygdala activate fear. He focused on the different kinds of neurons in the central nucleus, some on its lateral side and others on its medial side. Most importantly, he used very brief light pulses and specialized viruses to solve a major problem in contemporary neurobiology: when different kinds of neurons are next to each other, how is one kind influenced without at the same time influencing the other?

First, Anderson's lab members identified a chemical constituent of inhibitory neurons, the so-called PKC delta, on the lateral side of the central nucleus. Then they used their knowledge of the PKC delta gene to create a neuroanatomical marker and to show that those neurons send axons over to the medial side of the central nucleus. Those axons are thought to inhibit the medial neurons. Finally, after electrical recording experiments to verify their methodology, they suppressed neuronal activity in these PKC delta neurons and caused elevated levels of fear. They concluded that the lateral side of the central nucleus contains an "inhibitory microcircuit" that in turn regulates an inhibitory connection over to the medial neurons. Knock down that latter inhibition, and fear increases.

I am explaining brain mechanisms underlying fear for a serious purpose that leads to Chapter 5. The neuroanatomy, genetics, and neurochemistry are as complex as you would expect from studies of the mammalian brain, and these mechanisms cannot be described without a certain level of technical detail. But, in view of the precision and depth with which we understand brain mechanisms underlying fear-like behaviors (for mice, that often means "freezing"), it is easy to envision that scientists in the near future will be able to construct equivalent circuitries in robots.

Anderson also has ventured outside the amygdala to investigate another part of the ancient forebrain; the lateral septum. This receives inputs from the amygdala. Todd Anthony, working with Anderson, modified a particular set of lateral septum neurons that express a special receptor for CRF, mentioned above. Once modified, these neurons could be activated by brief light pulses sent into the septum even when other neurons nearby would not be activated. The bottom line is that activating these CRF-receiving neurons in the septum increased anxiety-like behaviors (avoidance of open spaces, etc.) when mice were stressed. Thus, the amygdala and its outputs, with an emphasis on the neuropeptide CRF, tell us how, in neuroanatomical and neurochemical terms, fear and stress feed into an anxious disposition.

Anderson and his team deciphered an elusive neurobiological mechanism. They related expression of specific genes in identified neurons of a mammalian brain to the causes of a global behavioral state.

Even more recently, Gregory Quirk at the University of Puerto Rico School of Medicine has shown how fear memories shift around in the brain as a function of the time since the fear training. At first, the basolateral amygdala and the prefrontal cortex are "hot" — extremely active. But days later the focus within the amygdala has switched to its central nucleus and a portion of the thalamus gets hot. It is not possible to discover this kind of neuroanatomical and neurophysiological transformation unless there is comprehensive understanding of the system. That understanding will assist in helping engineers to build emotional mimicry into robots.

Hormones. Hormones are also involved. Louis Muglia is a molecular endocrinologist and a pediatrician. Working with his team, then at Washington University in Saint Louis, he brought hormones associated with stress and fear into relation with the amygdaloid neuronal mechanisms just mentioned. One of these, cortisol (in animals, corticosterone), is a steroid hormone exquisitely responsive to stress and fear. It regulates gene expression in cells with its receptors in their cell nuclei, including cells in the amygdala. One of the neurochemicals it regulates is abbreviated to CRF (or CRH), which, like a tiny protein called a neuropeptide, has two separate roles: CRF directs the pituitary gland to send hormonal signals to the adrenal glands, thus to produce more cortisol; crucially, CRF acts in the brain itself to direct the organism's responses to fear and stress. In a clever series of experiments, Muglia and his team used viral vectors to knock out expression of cortisol's nuclear receptor in the central nucleus of the amygdala. The mice which received this treatment showed much less conditioned fear. Then, putting CRF back into the brain of such mice restored fear behavior. Muglia's team concluded that fear required cortisol signaling in the central nucleus of the amygdala with the consequence of CRF signaling then fear.

While the most discriminating analyses of amygdala neurons have been carried out in laboratory animals, the conclusions of such analyses should also apply to the human brain. For example, as mentioned in Chapter 2, Leslie Ungerleider at the National Institute of Mental Health and Ralph Adolphs at Caltech have shown the importance of the human amygdala for modulating behavioral responses to different types of faces.

In sum, from the details of the molecular genomics of forebrain neurons to the scanning of the human brain, neurobiologists are homing in on thoroughgoing explanations of what fear means in terms of brain mechanisms. The accomplishments of these laboratories will support the argument of Chapter 5, which is that the depth of our current and future understanding of mechanisms for emotional behaviors in the human brain will allow us to program equivalent mechanisms into robots in the near future. But what will robot fear-like behavior look like? Perhaps we could make a robot sweat, but that does not sound very smart. Instead, robots will anticipate threats to themselves and to those they serve with avoidance behaviors that prevent or reduce such dangers. As New York University Professor Joseph Ledoux would likely say, robots will exhibit "defensive responses" to threats.

Sex, Caring

Robots will certainly be able to reproduce themselves — in fact, they already have — but we do not know if, under any given circumstances, the logic of reproductive physiology in animals or humans will bear any resemblance to the regulation of robot reproduction. In my lab the genetics, neuroanatomy and neurophysiology of simple sexual acts have been worked out, and relate to the physiological side of what Freud called "libido." I mention this simply to demonstrate that neuroscientists are making great progress in understanding the intimate details of brain mechanisms for highly emotional behaviors.

Will robots be programmed to exhibit attraction to other robots and, if so, why? There may come a time when such attraction is useful. However, it is important to discuss what we know about animal brains when it comes to sex and caring, not only because of robot reproduction but also because, in humans, parental caring behavior extends from strictly reproductive behavior. Robots will care for us under some circumstances, and they may be programmed to care for each other as well.

In laboratory animals, sexual approaches are largely controlled by sex hormones: testosterone and its metabolites in males, estrogens and progesterone in females. Over a period of years, scientists worked out the entire set of cellular, circuitry, and molecular mechanisms for a sex behavior typical of females — a postural adjustment that is absolutely required for fertilization (Figure 4.5). Briefly, estrogenic hormone effects on hypothalamic neurons increase a hypothalamic output signal to the midbrain. Those midbrain neurons command a neural circuit that stretches from the sensory neurons of the spinal cord up to the midbrain and then back down to the spinal cord, thus to activate the motor neurons for the behavior.

Estrogens act in hypothalamic neurons by turning genes on. For example, consider this pair of genes, coding for oxytocin and its

Neural circuit.

Neural circuit for lordosis behavior.
From *Estrogens & Brain Function*.
(Springer-Verlag).

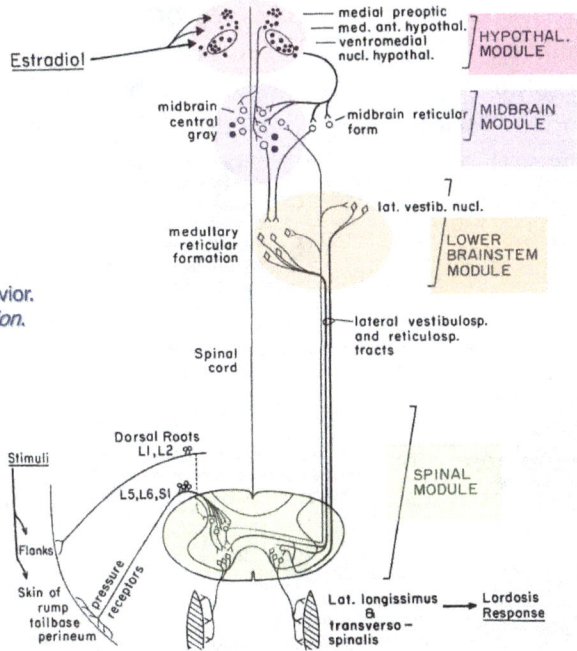

Figure 4.5 Mechanisms for sex behaviors in animal brains are well-understood. Sketched here is the circuit for the primary female reproductive behavior — the first circuit to be achieved for any vertebrate behavior. It comprises an estrogen-dependent signal coming from estrogen-binding neurons in the hypothalamus — a signal that facilitates activity in a spinal–midbrain–spinal neuronal look that regulates the behavior. (Adapted from Pfaff D, *Estrogens and Brain Function*, with permission.)

receptor: not only do estrogens turn on the gene for oxytocin, which fosters female mating behavior, but they also turn on the gene for the oxytocin receptor. These two genomic effects would have a multiplicative action to push female mating behavior forward at just the right time for appropriate reproduction.

If the gene for the primary estrogen receptor is knocked out, as we did some years ago, female reproductive behavior simply does not occur. And if it is knocked out in a specific set of neurons just in front of the hypothalamus, maternal behavior simply does not occur. Thus, we understand how expression of a specific gene is required for a train of behaviors that are very different from each other, but that are all essential for normal sex and parental caring.

The mechanisms that were worked out in simple laboratory mammals have by and large been conserved through evolution into the nonhuman primate and the human brain. Indeed, the pioneering psychoanalyst Sigmund Freud divided the concept of libido into two parts: the primitive physiological component and the more recently evolved cultural component. In my book *Drive*, I describe how we explained the primitive physiological contribution to libido. Now, Helen Fisher, a physical anthropologist who writes widely about the evolutionary underpinnings of human social behaviors, has teamed up with neuroanatomist Lucy Brown to show that neurons in dopaminergic reward-related systems and in subcortical action control systems light up with activity when the subject in the fMRI imagines his or her intensely loved long-term partner. In their words, "romantic love uses subcortical reward and motivation systems to focus on a specific individual," — the beloved. The essential point is that even this most personal and private of emotions has been subjected to systematic scientific inquiry.

Biologists with broad views of animal social behaviors emphasize that caring behaviors do not stop with caring for one's partner. Eminent Cambridge professor Barry Keverne not only talks about parental behaviors that depend on the same hormonal supports as lactation, but also extends his thinking to affiliative social behaviors in general. He describes a wide range of friendly, caring behaviors as being "emancipated" from hormonal controls. The brilliant primatologist Sarah Hrdy from the University of California at Davis, extends her analysis of normal maternal behavior in a different way. She shows how care for babies can be shared among females — she calls it "alloparenting" — generalizing from a narrowly described set of reflex acts to a generally helpful and friendly set of social behaviors.

Thus, a wide range of biological and brain-centered studies have told us much about sex and other forms of caring.

Anger, Aggression

Even robots could show anger-like behaviors. They could blatantly refuse to do a task that might harm them. They could complain when overworked. Understanding the dynamics of anger will help during the development of optimal robot–human interactions.

In the lab, we are almost always studying aggressive behavior as an objectively defined set of motor responses, not as an "emotional state." In the back of our minds we are thinking about feelings of anger, but the transition from an objectively measured behavior to a psychological state is not necessarily a transparent one. Therefore, I will talk about "aggression."

More than 50 years ago, Walter Hess, a physiologist working in Zurich, discovered that when he electrically stimulated neurons in certain parts of the hypothalamus or midbrain, this could cause cats to become angry or aggressive. He won the Nobel Prize for that work. John Flynn followed it up in an interesting way. A retired priest, Flynn became a professor of behavioral neuroscience at Yale and extended Hess' work by showing that stimulating in different parts of the hypothalamus caused different kinds of aggressive responses. Electrical pulses delivered to the medial part of the hypothalamus caused the cat to act enraged, while stimulation of the lateral hypothalamus led to what Flynn called a "quiet, stalking attack."

Hormones again come into play. We have known for two thousand years that eunuchs do not start street fights. In the modern neuroscience laboratory, Geert Devries, now at Georgia State, led the way in showing that testosterone effects on specific groups of neurons in the basal forebrain encourage aggression in male laboratory animals. Correspondingly, a preponderance of murders are perpetrated by young male humans (Figure 4.6). Devries and his colleagues pinpointed neurons that express the neuropeptide vasopressin as being important for androgenic stimulation of male aggression. Accordingly, when male laboratory animals are castrated, vasopressin levels are markedly reduced, and so is aggression. Put briefly, testosterone raises activity in neurons that promote aggression and reduces activity in neurons that inhibit aggression (Figure 4.7). Testosterone's most obvious route of action would be through its facilitation of the expression of specific genes in the brain (Figure 4.8).

Harrison Pope at Harvard Medical School was the first to draw a connection between testosterone and human aggression. But, if you

**DALY AND WILSON,
"UNIVERSAL" TREND:**

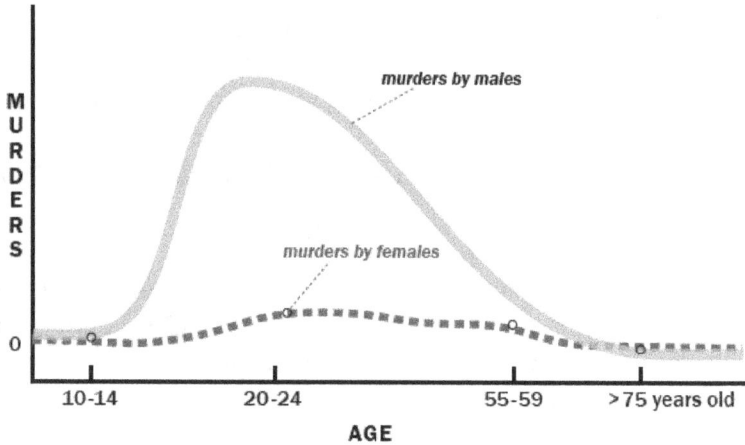

Figure 4.6   Murders of males by unrelated males are shown by Canadian psychologists Daly and Wilson to follow a lifetime curve that is similar to the lifetime curve of testosterone in the blood.

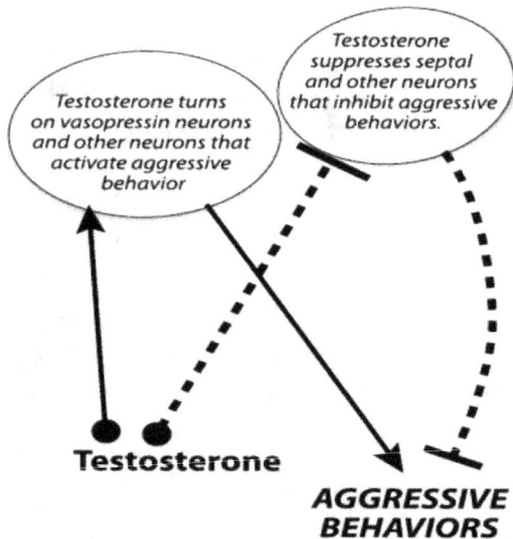

Figure 4.7   Testosterone facilitates gene expression and electrical activity in nerve cell groups that facilitate aggression; it also inhibits activity in nerve cell groups that inhibit aggression.

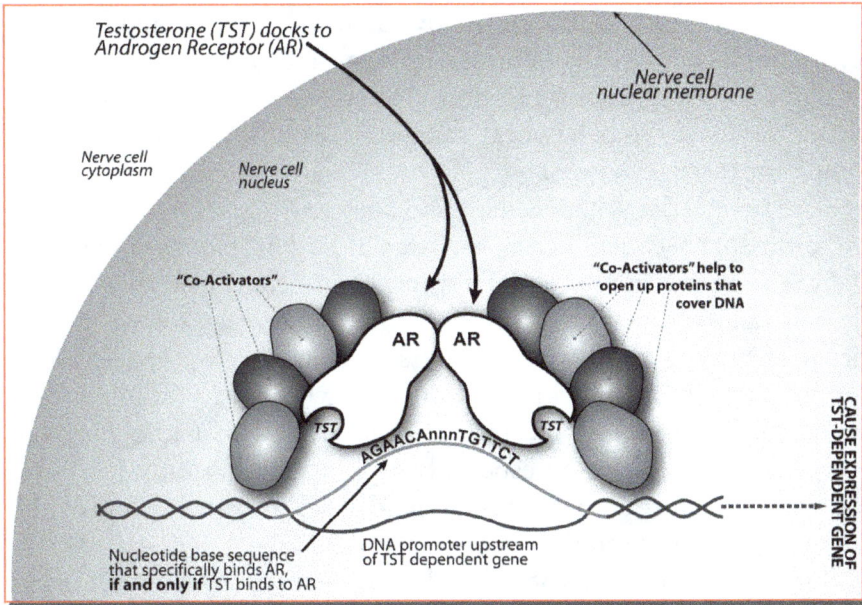

Figure 4.8 Testosterone (TST) can facilitate expression of genes in selected nerve cells following its binding to androgen receptors (AR). [Adapted from Pfaff, Goldman and Rapin, *Autism Research* (2011), with permission.]

reflect for a moment on the hypermuscular roles in various dramas or the real-life phenomenon of "roid rage," the scientific proof is obvious. Pope's work complements a huge set of studies on human aggressive behaviors summarized in Randy Nelson's book, *Biology of Aggression*. Loading up humans with alcohol or using drugs that deplete the neurotransmitter serotonin in the brain are sure to elevate aggression. Conversely, making serotonin more available at synapses in the brain decreases aggression.

Nelson's own laboratory's work on aggression was surprising because, instead of studying regular, conventional neurotransmitters or the kinds of hormonal influences mentioned above, he used a transmitter which is actually a gas — nitric oxide. When he knocked out the gene that produces nitric oxide, he produced animals that attack fast and attack often. Likewise, when he treated mice with a drug that inhibits that nitric-oxide-producing enzyme, those mice were extremely aggressive.

Genes. Years ago, my lab showed that we could make causal connections between the expression of specific genes and aggressive behaviors. Having discovered hormone receptors in the brain, I was excited to chart the behavioral consequences of knocking out genes coding for these receptors.

The connections between these genes and aggressive behaviors are different between males and females. For example, Sonoko Ogawa in my lab (now a professor in Japan) determined that knocking out the gene for an estrogen receptor *raised* aggressive behaviors by females, but *lowered* aggressive behaviors by males. Next, Masayoshi Nomura, working at Rockefeller University, studied a different nuclear hormone receptor and discovered that knocking out this gene in males caused especially high aggression only just after puberty; this maps onto the human tendencies for rates of murder by males to peak between the ages of 13 and 30.

Our work with gene knockouts leads directly to David Anderson's work, which has taken the molecular analysis of aggression in a new direction. First, Dayu Lin, then working with Anderson, used brief pulses of light transmitted to the hypothalamus through optical fibers to stimulate activity in ventromedial hypothalamic neurons that had been infected with a special viral vector that permits light to be translated into action potentials. Such "optogenetic" stimulation elicited a rapid onset of coordinated and directed attack toward an intruder. Optogenetic stimulation means that the hypothalamic neurons have been induced to express the gene ("genetic") a protein that turns a tiny light pulse delivered to the brain ("optic") into a series of electrical action potentials.

Then Anderson and his team went on to identify exactly which type of ventromedial hypothalamic neurons could carry the effect of optogenetic stimulation into attack behavior. By designing a viral vector dependent on the neuron with estrogen receptors, they could show that neurons expressing the gene for an estrogen receptor, when activated, caused the attack behavior. Nearby ventromedial hypothalamic neurons, not expressing estrogen receptors, did not have that effect. The more the optical stimulation onto these virus-carrying, receptor-expressing neurons, the more the attack.

In Anderson's words, "Recordings from the lateral corner of the ventromedial nucleus of the hypothalamus (VMHvl) during social interactions reveal overlapping but distinct neuronal subpopulations involved in fighting and mating." On the one hand, I am reminded of the thinking of Nobel Prize–winning ethologist Niko Tinbergen when he talked about "action-specific energies" — motivational forces that might be directed toward one natural behavior sequence of another, for example, sex versus aggression. On the other hand, males of several species must defend their territories. According to behavioral neuroscientist Jaak Panksepp, "In virtually all mammals, male sexuality requires an assertive attitude, so that male sexuality and aggressiveness normally go together."

In sum, aggression, one type of animal behavior, and anger, one type of human emotion, serve valid and enduring purposes. Food must be fought for; territories must be maintained. From an evolutionary point of view and from the laboratory bench, we understand much more about this cluster of emotions than we did 50 years ago. The dynamics of incipient aggression, and its prevention, will play into human–robot interactions, as argued in Chapter 5 and further envisioned in Chapter 6.

Disgust

Since this chapter provides an overview of our tremendous breadth and depth of knowledge about emotional mechanisms, I must introduce a less-than-pleasant topic: disgust. Robots indeed may "get disgusted" in that their behaviors may cause them to avoid what they have been programmed to treat as "disgusting" — but that is not the case that I make here. Instead, I simply want to introduce a surprising set of investigations about this little-studied emotion, here connected with a physiological reaction evolutionarily designed to protect us.

As my book *The Altruistic Brain* notes, Paul Rozin, a professor at the University of Pennsylvania, has taken a part of the forebrain long associated with simple sensory appreciation and tied those neurons to an unpleasant emotional state — disgust. Nerve cells in a deep, hidden part of our cerebral cortices, the insula, had long been known for their responses to taste and for our sense of our overall body condition, for example, in our viscera. It now turns out that neurons in the insula also are sensitive to the moral qualities of our social acts. As a result, Rozin coined the phrase "from oral to moral," describing roles for neurons in the insula, from disgusting taste to immoral, disgusting behavior.

Happiness and Unhappiness

What we want to know about a robot's "feelings" (next chapter) is whether or not the robot is happy in its work. That is to say, the robot may be functioning perfectly — everything going smoothly — or, for example, it may be frustrated in its work by mechanical problems in its gripper arm or by sensing that its internal temperature is not within an acceptable range. In these latter examples, it may sense that it will not be able to fulfill all of its command, within designated limits. Those problems are likely to produce what could be called an "unhappy robot."

Happiness and its opposite state have not received nearly as much attention from neurobiologists as you might have imagined. But there is one exception. Thus, scholars have work to do to define exactly what happiness means in terms of human brain mechanisms. Engineers must work to prevent "robot unhappiness," whether that means answering error reports or robotic "chronic fatigue."

However, Richard Davidson, a professor of psychology at the University of Wisconsin, has made advances in this regard. More than any other scientist, he has taken on "mental states" as serious subjects for neuroanatomical inquiry. As he recalls in *The Emotional Life of Your Brain*, he was inspired by an old paper by an Italian neurologist at the University of Perugia — Guido Gainotti. Gainotti observed that patients with severe damage to the front part of the brain on the *left* side cried a lot irrespective of events of the moment, and also displayed symptoms of depression. In contrast, damage on the *right* side produced what Davidson called "pathological laughter."

Davidson and his team, who are interested in the development of emotional dispositions, then studied ten-month-old infants by noninvasively recording electrical activity from the surfaces of their little heads. Now, what would you predict from the earlier Italian study? If damage to the right side releases laughter, then the right side of their frontal cortex specializes in the opposite: negative emotions. For the left side just the opposite occurs: positive emotions. He confirmed this. In his prestigious *Science* paper, he reported that when the babies saw happy scenes they smiled and the electrical activity of their left frontal cortex soared. The opposite was true for the right frontal cortex.

These left–right differences may seem surprising, given the following perspective. Try to recall the sketches of sensory systems in Chapter 2 and motor systems in Chapter 3. The left–right differences in those systems make good sense. The left visual field is coded into one side of the visual cortex; the right one is coded into the other. Likewise, motor cortical neurons on the left side initiate movement by muscles on the right sides of our bodies, and the reverse is also true. But the Davidson team's findings opened up the field with respect to the sidedness of emotional regulation. Indeed, Sara Schaafsma, a postdoctoral researcher in my lab, found a left–right difference in gene expression in the hippocampus in relation to emotional behaviors. It is therefore likely that in the very near future the concept of "happiness" will be reified in the electrical and chemical activities of well-defined circuits in lab animal brains and in humans.

Arousal

As mentioned above, underlying the expression of all emotions is a primitive force that I have called "generalized brain arousal" (GA). GA is produced by the activity of neurons in the core of our brainstem, starting just above the spinal cord and going all the way up past the midbrain into the thalamus and into the basal forebrain. Many neurotransmitters participate: norepinephrine, dopamine, serotonin, histamine, and acetylcholine. For example, when we took old-style allergy medicines, they made us sleepy. That is because their antihistamine chemistry blocked H1 receptors, thus blocking the arousing actions of histamine-producing neurons. Hormones participate in GA as well. Too much thyroid hormone and we are hyperalert. Too little and we are dull.

We now know exactly how arousal-causing transmitters impact individual emotional systems. For sex, the arousal transmitters histamine and norepinephrine help to turn on electrical activity in the hypothalamic neurons responsible for initiating sexual acts. For fear, workers at the State Key Laboratory for Medical Neurobiology in Shanghai, China, have shown how a specific neurochemical responsible for regulating responses to arousal-causing transmitters actually changes its position within amygdala neurons in response to fear conditioning. In both cases, emotional "chemistry" is being boiled down to neurochemistry.

Thus, while individual feelings account for the unique character of every emotional state, GA accounts for the intensity and strength of its expression. Brain mechanisms for every emotion have a dual quality: those that produce an aroused brain, and those peculiar to each individual feeling (see Figures 4.3 and 4.4).

How will an aroused robot act compared to a nonaroused robot? First of all, the aroused robot will be ready to work. Just as important, from the point of view of this chapter, an aroused robot will be capable of expressing the types of emotion-like behaviors that will smoothen its interactions with humans in tricky situations that require empathy or at least some social coordination. Designing arousal systems in robots will be easy. They will act something like sophisticated, highly articulated power supplies.

Sex Differences?

People often talk about sex differences in behavior without much attention to the subtleties of sexual differentiation of the brain or the many choices of gender roles. Sex differences in the brain depend on myriad genetic, chemical, and physical mechanisms, while gender roles depend on seemingly infinite sets of social conditions.

Sex differences in human emotional functions show up most readily when doctors are studying emotional disorders (see *Frontiers in*

Neuroendocrinology). Psychiatrists have long known that women suffer from depressive and anxiety disorders much more frequently than men. On the other hand, crimes of violence ("acting out" and worse) and symptoms of autism show up more frequently in boys and men. Since we will want to ensure that robots never act violently except when designed to be soldiers, robot self-defense programs will have to be designed with care. Supermoral robot leaders may be necessary. As mentioned in Chapter 1, Ronald C. Arkin, a computer engineering leader at Georgia Tech, has argued forcefully that robots can be programmed to behave exactly as dictated by civilized rules of war. For example, if he is correct, robot soldiers would not have committed heinous acts that American soldiers and Blackwell consultants committed in a small number of incidents in the second Iraq war.

Where might these sex differences come from? The genetics are much more complicated than just the Y chromosome and, obviously, adverse environmental influences can work on unfortunate disease states in myriad ways. As reviewed by the University of Michigan's Jill Becker and her colleagues, brain mechanisms for the interactions between genes and the environment include hormone receptors, particularities of gene expression, neuroanatomy, autonomic physiology, and subtle differences in developmental trajectories of behavior.

In sum, sex differences are a complicated but undeniable feature of emotional life. Since humans will constitute important components of robots' environments, and since those environments feature human emotions, governing our own behaviors (whether sexually differentiated or not) to remain appropriate to robots' individual differences will surely prove necessary.

We do not know yet whether we are going to have gendered robots. Would doing that improve the range and flexibility of human–robot interactions? Indeed, would it allow programmers to take advantage of what we already know about male–female relations in the workplace? Certainly, it will be possible to build reproductive systems into robots — to engineer systems by which robots replicate themselves — but it remains an open question as to how much our knowledge of human reproductive hormones will aid that endeavor.

Bottom Line

In the words of Nobel Prize–winning neuroscientist Eric Kandel in his book *The Age of Insight*, the modern neuroscience of emotion grew out of evolutionary thought, psychoanalysis, and solid physiology, but "current thinking about emotion has also been influenced by cognitive psychology and by the developmental of more sensitive measurements of physiological changes within the body." The preface states: "The central challenge of science in the 21st Century is to understand the human mind in biological terms." In this chapter we have examined remarkable progress in approaching that goal with respect to the neurobiology of emotion.

The neuroanatomy and the physiology of emotion regulation can be contrasted with sensory and motor systems. In the latter two systems, point-to-point connectivity representing the three-dimensional geometry of our bodies and our environments makes a lot of sense. Those neural systems permit well-organized specific responses to stimuli that are discrete in time and space. Emotional systems are different. Inputs of affective importance bomb into certain primitive forebrain neuronal groups like the amygdala from all over and their effects can last for a long time. Subsequently, outputs radiate from these neuronal groups to regulate a wide variety of behavioral, hormonal, and physiological systems, often for a long time.

The accomplishments of neurobiologists in this area and the intensity of ongoing work convince me that our understanding of emotional systems in animals and humans will usefully inform our programming of social robots. Chapter 5 will discuss that point systematically.

Further Reading

Anthony TE, Dee N, Bernard A, Lerchner W, Heintz N, Anderson DJ (2014) Control of stress-induced persistent anxiety by an extra-amygdala septohypothalamic circuit. *Cell* **156**: 522–536.

Becker J *et al.* (eds.) (2008) *Sex Differences in the Brain.* Oxford University Press, New York.

Damasio A (1998) Emotion in the perspective of an integrated nervous system. *Brain Res. Rev.* **26**: 83–83.

Damasio A (1999) *The Feeling of What Happens*. Harcourt Brace, New York.

Davidson R, Begley S (2012) *The Emotional Life of Your Brain*. Hudson St. Press (Penguin), New York.

Debiec J, LeDoux J (2009) The amygdala and the neural pathways of fear. In: Shirmani P *et al.* (eds.), *Post-traumatic Stress Disorder*. Humana, Springer, Heidelberg.

Haubensak W *et al.* (2010) Genetic dissection of an amygdala microcircuit that gates conditioned fear. *Nature* **468**: 270–275.

Kandel E (2012) *The Age of Insight*. Random House, New York.

Kolber BJ, Roberts MS, Howell MP, Wozniak DF, Sands MS, Muglia LJ (2008) Central amygdala glucocorticoid receptor action promotes fear-associated CRH activation and conditioning. *Proc Natl Acad Sci USA* **105**(33): 12004–12009.

LeDoux J (1996) *The Emotional Brain*. Simon and Schuster, New York.

Lee HS, Kim DW, Remedios R, Anthony TE, Chang A, Madisen L, Zeng HK, Anderson DJ (2011) Scalable control of mounting and attack by Esrl neurons in the ventromedial hypothalamus. *Nature* **470**: 221–225.

Lin D, Boyle MP, Dollar P, Lee HS, Perona P, Lein ES, Anderson DJ (2014) Functional identification of an aggression locus in the mouse hypothalamus. *Nature* **509**: 627–631.

Nelson R (ed.) (2006) *Biology of Aggression*. Oxford University Press, Oxford.

Ogawa S, Nomura M, Choleris E, Pfaff D (2006) In: Nelson (ed.) (*op. cit.*).

Panksepp J (1998) *Affective Neuroscience*. Oxford University Press, New York.

Pfaff D (1999) *Drive*. MIT Press, Cambridge, MA.

Pfaff D (2006) *Brain Arousal and Information Theory*. Harvard University Press, Cambridge, MA.

Pfaff D (2014). *Altruistic Brain Theory*. Oxford University Press, New York.

Pfaff D, Young L (2014) *Frontiers in Neuroendocrinology: Sex Differences in Neurological and Psychiatric Disorders*. Elsevier, Amsterdam.

Quirk G (2012) Fear. In: Pfaff D (ed.), *Neuroscience in the 21st Century*. Springer, Heidelberg.

Schaafsma S, Pfaff D (2014) Etiologies underlying sex differences in autism spectrum disorders. *Front Neuroendocrinol* **35**: 255–272.

Sears RM, Fink S, deLecea L, LeDoux J (2013) Orexin/hypocretin system modulates amygdala-dependent threat learning through the locus coeruleus. *Proc Natl Acad Sci USA* **110**: 20260–20265.

Smith E, Kosslyn S (2007) *Cognitive Psychology*. Pearson/Prentice Hall, Saddle River, NJ.

CHAPTER 5

Regulation and Display of "Emotions" by Robots

This chapter will explain how robots can display a type of affective behavior that we can perceive as, or analogize to emotion. Having examined how human systems work, we will see that robotic "emotions" will perform functions comparable to those of human emotions, signaling the robots' self-perceived state. For example, robots will signal concerns about their continued working order ("health") or about their environment ("threats" from other robots). I will argue that robots should be capable of communicating their capacity to function in settings where we need to know how they are functioning — just as humans can communicate through emotions that convey their own self-assessments.

Taking off from the neurobiology in Chapter 4, I will argue that robotic displays of *apparent* emotion will, eventually, be virtually indistinguishable from human emotional responses. A tear will look like a tear. A grin will look like a grin. Then I will ask: How do we recognize human emotions anyway, since the capacity to read signals — whether human or robotic — is basic to communication and indeed to coexistence? Ultimately, I will consider whether humans' ability to read each other's emotional signals "translates" into a capacity to read signals coming from robots. At first, in the very near future, it is likely that robotic signals, no matter how sophisticated the robot, will be less subtle than the emotional signals coming from humans, if for no other reason than that the range of human signals will be far greater and far less grounded in some utilitarian objective. But the five-part argument below suggests that the subtlety and appropriateness of emotion-like expressions by robots will increase markedly over the years.

The key phrase throughout this chapter will be *as though*, since even as we explore the correlation between human and robotic emotion — which, of course, can never be exact — we must keep in mind that any such correlation is a *logical construct* in aid of building robots suited to working among humans. That is to say, we must think of robots *as though* they had emotions that would aid them in communicating with us. These so-called emotions will be built into robots based on (though not always following) principles that we understand from our study of human emotion.

"Emotional" Behaviors of Robots

Robots can be programmed to behave as though they had human-like emotions. Consider five simple steps to this conclusion:

(1) *Human emotions — feelings — are produced by mechanisms in the human brain.*

Chapter 4 showed that several lines of experimental evidence are converging that describe how emotions work. In the earliest days of scientific inquiry, conclusions depended on straightforward neuroanatomy and neurophysiology in laboratory animal brains coupled with measures of animal behaviors such as fear, sex, and aggression. Human subject studies depended on medical summaries of patients with brain damage. Today, laboratory scientists employ the tools of molecular biology and genetics, and have access not only to functional magnetic resonance imaging (fMRI) but also to multiple techniques built on fMRI: axon tract tracing, coherence among brain regions, and others.

As a result, we are coming to understand how, in cellular and chemical detail, emotional behaviors and the feelings associated with them are produced by neurons in brain areas such as the hypothalamus, septum, insula, and amygdala. Further, we are beginning to see how those feelings are regulated by neurons in our frontal cortex.

While my lab has focused on behaviors and emotions that serve sexual relations and that depend on the amygdala, work by Ralph Adolphs' lab at the California Institute of Technology, and many other labs, highlights the roles played by nerve cells in the amygdala. Such laboratory experiments are described in Chapter 4, and provide a tiny sampling of the avalanche of new scientific findings on mechanisms for emotion.

So point 1 of this argument is: human emotions — feelings — are produced by mechanisms in the human brain.

(2) *These mechanisms can be discovered.*

Likewise, the discoveries summarized in Chapter 4 prove point 2, though for brevity's sake they vastly under-reported our progress. Take a look at substantial books like Joseph LeDoux's about fear, my book about sex, and Randy Nelson's about aggression. Neuron by neuron, transmitter by transmitter, hormone by hormone, the mysteries of emotional regulation by the brain are rapidly being figured out.

(3) *Once discovered, these mechanisms can be precisely imitated by computer programs.*

For the last 60 or 70 years, one could say with confidence that if a complex system's performance could be described accurately and in detail, then it could be imitated by a newly designed electronic or digital system that copied it. Even in the 1960s, when I took a course at MIT with the great teacher Michael Dertouzos about "finite state automata," it was obvious that no upper limit would be placed on the sophistication of systems.

However, here we are dealing with a much simpler issue than those which have occupied philosophers and computer scientists for almost a century. For example, the great applied mathematician and computer scientist Alan Turing asked: "Can machines think?" He stated: "The nervous system is certainly not a discrete-state machine. A small error in the information about the size of a nervous impulse impinging on a neuron may, make a large difference to the size of the outgoing impulse. It may be argued that, this being so, one cannot expect to be able to mimic the behavior of the nervous system with a discrete-state system."

But, for my argument, we are not looking for the robot's computers to mimic, exactly, what a human's emotional nervous system does. Within this limited domain, we need only a reasonable approximation of affective behavioral outputs for limited times. This will prove to be relatively easy as neuroscience surges forward, combined with sophisticated robot design.

In *Psychologism and Behaviorism*, philosopher Ned Block at New York University envisions machines that "simulate our neurophysiology," and he proposes that systems that this "could have actual and potential behavior *typical* of familiar intelligent beings." Both of those things will happen. However, I do not propose what John Searle, a philosopher at U Cal Berkeley, calls "strong Artificial Intelligence," claiming that "the appropriately programmed computer actually *is* a mind." Instead, I support his "weak Artificial Intelligence": future robots designed for social interaction will be able to produce behaviors that *look and sound as though the computer had* a mentality that embraces emotions.

That is to say, in Searle's and Block's terms, I am only looking for a weak equivalence between the human's emotional expressions and the robot's behaviors. Simply, given the same sensory inputs, their behavioral outputs should be virtually indistinguishable. This requirement does not demand that every emotionally relevant state of the human's nervous system be matched by a corresponding state in the robot's computers. Instead, for any given robot's range of responsibilities to any given set of human partners, affective and communicative states should correspond well enough to get the job done.

If step 3 of the argument sounds extreme, remember that the robots we are discussing are not "all-purpose entities" like human beings. Instead, they will be designed for particular uses. That facilitates programming, and makes step 3 not only feasible but relatively straightforward. Accordingly, philosophical arguments should not discourage us from thinking that, once discovered, at least some emotion-producing mechanisms in the human brain can be precisely imitated by computer programs. Philosophical arguments are at a level of abstraction that does not come to terms with the intersection of modern neuroscience and computer science.

(4) *Such computer programs can be used to regulate all relevant aspects of robot behavior.*

Since we are talking only about robots that interact with human beings, the programs mentioned in step 3 will concentrate on social behavior, that is, they will regulate how robots behave toward us. This will include such behavior that appears to us to be emotional.

(5) *As a result, robots can behave perfectly as though they were having the human emotion imitated.*

That emotion as imitated may seem appropriate to the human companion, or inappropriate, but in either case, the human will recognize it as "emotional."

In light of this five-step argument, think back to the discussion in Chapter 4 that emotional inputs to the CNS report on the state of the organism. Thus, some might say that if robots are designed properly, they will be able to tell us directly and literally all that we need to know about them, without any intermediating gestures on their part or intuitions on ours. Robots will use plain language, or the electronic equivalent of clearly readable signals. Machines can do this already, as when an alarm goes off alerting managers of a nuclear plant that a vessel has overheated and is in danger of exploding.

The counter-argument would say that we do not need such things as "unhappy" robots, but only a robot that conveys in specific terms that it is currently or in danger of becoming inadequate for its job. Then the task for robot engineers will be to ensure that the robot has the necessary linguistic or electronic means to signal — that is, communicate to other robots or to humans — that it is failing and is in need of attention. Correspondingly, there need not be any such thing as a "happy" robot, but only a robot that has not diagnosed any problems in its performance and emits no signals to that effect. The best engineers will create the most acutely sensitive robots, ones that (a) constantly monitor their internal and external conditions in terms of how these affect the robots' performance, and (b) most quickly, comprehensively, and directly communicate to humans when some change in the robots' condition needs attention.

All of this counter-argument sounds great for engineers, but for all the rest of us, interacting with thousands of different kinds of sociable robots in thousands of situations, understandable affective language will go a long way toward making these situations productive. The five points above prove that such situations are possible. That is to say, there is no rule that we *must* respond to robots as though they had emotions. Instead, we now understand that we *can* in the near future do so.

In the future, robots will be able to analyze their conditions and speak about them insofar as they are programmed to preserve themselves and optimize their serviceability. If a robot says "My left arm is seriously overheated and my fingers could melt," we might rush to its aid. Neuroscientists and computer circuitry designers might respond to a hyperliteral robot in a hyperliteral way. But, for smoothening all human–robot interactions in a work-a-day world, emotional resonance helps a lot.

Progress to Report

Robot engineers have been busy. Consider first the recognition of emotional phrases.

In the words of Shiqing Zhang in the School of Physics and Electronic Engineering at Taizhou University in China, "Given the non-linear manifold structure of speech data, the traditional linear PCA method based on the linear assumption of feature data cannot effectively handle such nonlinear speech data." What this means is that Zhang's machines break up human speech into chunks of voice activity that do not include pauses longer than 200 milliseconds. Within each chunk the system recognizes voice pitch, sound intensity, and sound duration. Voice quality measures include spectral energy distribution, harmonics-to-noise ratio, pitch variability, and amplitude variability. Then, statistical methods are used to help the machine through "learning by data examples." Among a large set of emotional phrases including anger, joy, sadness, disgust, and fear, the system could claim an emotion recognition accuracy in the 70–80% range. Anger was the most easily recognized, at 90%.

Robots can make emotional expressions too. At MIT the robot called Nexi uses neck movements, gaze, eyebrows, etc. to give the possibility of emotional expression. And consider the Kansei robot, in Japan, which is supposed to make more than 30 different facial expressions. According to *Technology Review*, the word "*sushi*" elicits expressions of enjoyment.

Ronald Arkin's teams at Georgia Tech have already worked on the expression of nonverbal emotion-like behaviors by robots which can be recognized by human observers. His software — units that try to deal with "traits, attitudes, moods, and emotions" (TAME) and that feed into his open-source MissionLab software — deals with movements of the body as a whole, as well as gestures, a variety of postures, regulation of distance from the partner, and nonlinguistic voice qualities such pitch variation and voice volume. Across a lot of robot displays that were supposed to convey the ideas of various affective or mood states, some were much more successful than others. The best were that 85% of the human observers recognized the robot displays intended to denote "joy," and 81% "fear."

I predict that accomplishments like those cited here constitute the beginning of a tidal wave of demonstrations that robots can behave as though they recognized emotions and can display emotion-like behaviors.

How Do We Come to Believe Another's "Emotional Language," Robotic or Human?

If robots are going to behave as though they had emotions, and if we have para-emotional exchanges with them, they certainly will have to know how to respond to and emit typically emotional words. How will they learn to do that? Indeed, how do humans learn the meanings of emotional words?

When putting these questions to psycholinguists, they frequently responded with "We just don't know." For example, one expert explained that how the young child relates the word "sad" to the actual feeling expressed in the sentence "John is sad" is not clear.

Does the surrounding grammatical and semantic information help? The experts are uncertain.

Lots of theoretical approaches to word learning have been proposed, but one of them likely has the most sense: "statistical learning." According to this theory, children notice and learn the quantitative properties of the language they hear, and use statistical regularities to construct categories of words. One limited application of this approach is useful here: the categorization of words as "emotional words" and the subsequent distinctions among them.

For example, a leader in the field of statistical learning, Jenny Saffran, Professor of Brain and Cognitive Sciences at the University of Rochester, has written that "a fundamental task of language acquisition, segmentation of words from fluent speech, can be accomplished by 8-month-old infants based solely on the statistical relationships between neighboring speech sounds. Moreover, this word segmentation was based on statistical learning from only two minutes of exposure, suggesting that infants have access to a powerful mechanism for the computation of statistical properties of the language input." Saffran's data show that "a powerful learning mechanism capable of extracting statistical information from fluent speech is available early in development." Below, I propose that this powerful learning mechanism serves to decipher the emotional references of words. Children do it and robots could.

But, still, why should we want robots to understand our emotional words and gestures? Here's why. Robots will be performing an ever-greater number of specific tasks, and humans will depend on those tasks. Humans will therefore need to interact with robots and will need to know that the robots can react to human expressions. Further, if robots can use emotional words properly, we will understand how they are reacting to their task. For example, if a robot is about to explode because one of its circuits is overheated, it should be able to convey that it is extremely uncomfortable and on the verge of dysfunction. Such "feelings" will need to be specific and recognizable, so as to enable humans to act in a useful way.

So, taking Saffran's basic approach and adding details, how will robots learn emotional words? And how will we apprehend apparent

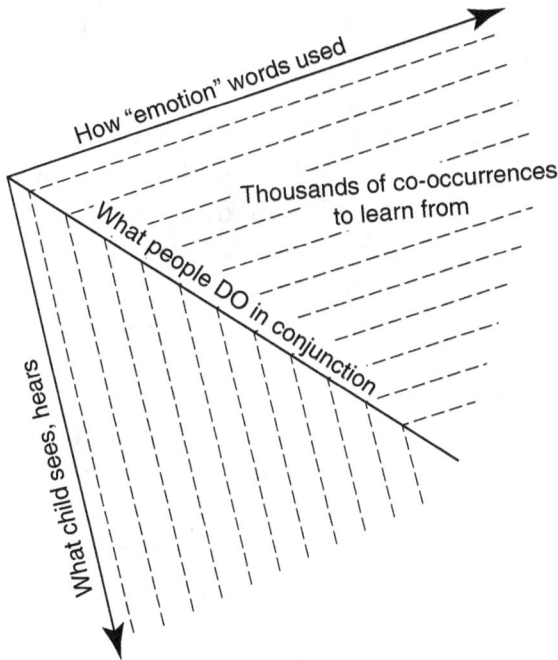

Figure 5.1   A three-dimensional sketch to illustrate the following idea: applying Jenny Saffran's statistical learning approach, children could learn to use emotion words appropriately from observing thousands of correlations: *word usage/see expressions and emotions/see emotional behaviors.* Robots may be able to learn the same.

feelings expressed by robots if they can learn to use affective language? Well, how do children learn language? And how do we infer the feelings of other humans when they are not employing obvious signifiers, such as laughing or crying?

According to my application of statistical learning, during development we build up, establish, and then employ huge mental tables of *correlations* between how we feel and the words we are taught to use for expressing those feelings (Figure 5.1). Similarly, we build up years of experience noticing feelings of others that we compare to words of still others when those others are obviously expressing emotions. On the expressive side, robots can be programmed to make laughing, crying, and other emotionally expressive sounds identical to those of humans. As a result of these correlations and sounds, robots will be able to be seen (and heard) as machines that *seem* to have feelings. Robots will express "emotions" appropriate to the situation and will use "feeling" words almost identically to humans. And they will be able to respond appropriately to our emotional words.

To put the same point another way, thousands of trials in early life allow us to construct a kind of three-dimensional correlation matrix (Figure 5.1), in which Dimension 1 is what the child sees and hears in a given moment; Dimension 2 constitutes what the emotional word user is doing at that time (including but not limited to his facial expression, crying, and laughing); and Dimension 3, in conjunction with Dimensions 1 and 2, *how emotional words are used* during that episode. After such a large amount of experience during normal human development, we act on the assumption that, in the usual case, "feeling words" report the feelings that we would experience ourselves. There is no reason why robots cannot benefit from the same learning processes as small children in learning emotional communication.

Yale professor Paul Bloom emphasizes other factors in language acquisition in *How Children Learn the Meanings of Words*. He criticizes the view of word learning as the easiest part of language development, and argues that "children's learning of words, even the simplest names for things, requires rich mental capacities — conceptual, social, and linguistic — that interact in complicated ways." Bloom urges us not to oversimplify. In addition to the "developmental" factors that influence exactly when children can learn words and how fast they can learn them, he emphasizes "experiential factors": "as you learn more of a language, you gain access to experiential information relevant to word learning." This is a positively accelerated process. As pictured in Figure 5.1, Bloom talks about learning words through nonsyntactic context. Thus, "the best way to learn a word through context is by hearing it used in a conversation with another person." Children do — and robots could — *selectively encode* (distinguish relevant from irrelevant information when learning the word); *selectively combine* (combine contextual cues to narrow to a workable meaning of the word) and *selectively compare* the new linguistic information to background knowledge. In sum, I take Bloom's points of view not to replace the straightforward explanations of Saffran and Figure 5.1, but to supplement them.

Thus, not only will programmers be able to construct emotion-like circuitry in robots, but also robots will talk about their apparent feelings in a convincing manner. We will be able to treat robots *as though* they had feelings.

Recognizing Another's Emotional and Social Intentions

An important feature of human emotional discourse occurs when one of us can sense another's feelings or intentions. Robots will be able to do that too.

During normal social exchanges, humans can often guess the assumptions and intentions of those they are watching or listening to. Scientists who study such abilities often group them under the phrase "theory of mind." Sara Schaafsma, a researcher in my lab at Rockefeller University, has criticized the use of this phrase, because it has come to encompass so many different human capacities that its use has become vague. She maintains that the phrase should be broken down into its separate components and then reconceptualized as an entire family of social behaviors. Nevertheless, no adequate substitute has yet been proposed. At the same time, we now know something of the nerve cells underlying a human's ability to exhibit theory of mind and its development in children, and we have ideas about its appearance in other animals.

Consider the severe upper limit on how complicated theory of mind could be. Infants can perform as though they understood intention and recognized good social behavior. I visited the lab of Karen Wynn, Professor of Developmental Psychology at Yale, and saw how she works. Each baby comes into the lab, sits on its mother's lap, and watches a puppet show designed to elicit its appreciation of prosocial versus antisocial behaviors. The "good puppet" (say, in a yellow jacket) helps a green puppet get something out of a box. The "bad puppet" (say, in a red jacket) not only obstructs the green puppet but hurts its hand. According to Wynn's data, infants as early as three months of age could see the difference between the kind puppet and the mean puppet, reaching out to touch the yellow-jacketed puppet when presented.

Likewise, Michael Tomasello, Co-director of the Max Planck Institute of Evolutionary Anthropology in Leipzig, Germany, did an experiment in which there was a play area marked by a blanket on the floor where a child and an experimenter used a sponge in the usual way, pretending to clean up spills and wash dishes. Then the adult brought the child and the sponge to another play area, a table, where a new game was played. A puppet then came in and used the sponge incorrectly, wiping things up inappropriately. Notably, about two-thirds of the three-year-olds would spontaneously protest, saying things like "No, you are not allowed to clean up here." Such experiments demonstrated that even toddlers quickly understand and follow new social norms. How hard, then, could this be? Not very hard, and easy to program.

In somewhat older children, 12–15 months, Jessica Sommerville and her colleagues at the University of Washington used an experimental paradigm called "violation of expectation." Her infants "watched an interaction between a distributor and two recipients," with the experimenters working on the idea, previously demonstrated, that the infants had already developed a sense of fairness. "An actor distributed crackers to two recipients, resulting in either equal or unequal portions." Experimenters found that the infants would pay more attention when the crackers were distributed unequally, that is, when the infants' expectation of fairness was violated. The infants could put themselves in the place of the cracker recipients and know what was deserved.

Some experimental psychologists might choose to interpret Wynn's, Tomasello's, and Somerville's data some other way, but it seems to me that neural systems as elementary as those of a very young infant can exhibit theory of mind. I think robots will also be able to.

Evolutionary biologists have learned to demonstrate theory-of-mind types of behavior in lower animals. You may not be surprised to read that chimpanzees show impressive and intimate social behaviors. Primatologist Tetsuro Matsuzawa at the University of Kyoto maintains a large laboratory for studies of chimpanzee behavior and also observes chimpanzees in the wild. He says that at early developmental stages chimpanzee and human behaviors resemble each other. The adult resonates with the baby chimpanzee in a striking manner when the baby is learning to use rocks to crack nuts for food. That learning process would not work if the adult did not intuit what the baby knows and what the baby does not know. Likewise, studying dolphins and elephants, Diana Reiss of Hunter College at the City University of New York finds evidence that these animals can distinguish self from other. While no one is trying to say that these animals represent the minds of their conspecifics (members of the same species) in the way that humans can, the scope of abilities that are considered uniquely human is shrinking steadily.

As scientists move forward with investigations of emotional and social behaviors and their brain mechanisms in humans, what about robots? Ehud Sharlin is a computer scientist at the University of Calgary. He has performed experiments demonstrating that human subjects readily form relationships with inanimate objects, especially ones capable of movement and designed to elicit social responses. In turn, the ubiquity of theory-of-mind phenomena from early human development and in lower animals suggests that we will be able to engineer behavior control circuitry that accomplishes the same social discriminations as chimpanzees and dolphins can.

We know already what social cues robots will react to. They will pay attention to the whites of the eyes, as mentioned above from the work of Ralph Adolphs. But they also will attend to body cues. For example, Hillel Aviezer led an international team whose subjects examined reactions of athletes when they had won or lost a tennis match. In this study, body position was more effective than facial expression in permitting the subjects to know if the athlete's feeling expressed a positive or negative emotion. All emotions are expressed as movement, and robots will be trained to recognize these movements and respond appropriately. Hundreds if not thousands of studies like these will tell robot behavioral engineers exactly what they need to know.

In summary, humans readily interact with robots, and robots will know how to respond. In terms of the "Turing test" mentioned in Chapter 1, there will be myriad times when a human will not know whether a robot or another human is responding. There is money to be made in this field. The company Affectiva, founded at the MIT Media Lab by Professor Rosalind Picard and her colleagues, not only develops hardware to measure autonomic and emotional responses, but also develops software by which robots could respond to human emotions. Picard's new spinoff project Empatica, an "affective computing company," has developed a wristband, essentially a watch-like computer that measures emotionally relevant physiological dimensions such as heart rate, heart rate variability, arousal, excitement, temperature and motion. Altogether, from the work of labs like mine, Joseph LeDoux's, Randy Nelson's, or David Anderson's (see Chapter 4) and companies like Picard's, the neuroscience of emotion has emerged from the dark ages and has become both exciting and profitable.

Outlook

Robots are taking over jobs that humans used to do, as we will worry about in Chapter 8. But we will still need to monitor them and work alongside them. That is to say, robots are becoming part of communities in the workplace. Thus, we will need to "understand" them. Note that, at first, they may be better at picking up signals from us than we are at picking up signals from them. This disparity itself may pose a problem. Such a disparity leads to the question of how we can engineer robots so that they are more "forthcoming," more able to convey complex states that we identify as feelings. However, fast this field develops — and it will develop — the questions posed in this chapter have registered in the public's consciousness. In addition to the movie *Her*, referred to in the Introduction, I was struck by the movie *Robot and Frank*, in which the aging guy played by Frank Langella expressed real and convincing feelings for his robot. Emotional relations with robots are coming. It is just a matter of time.

Further Reading

Adolphs R (2010) What does the amygdala contribute to social cognition? *Ann NY Acad Sci* 1191: 42–55.

Aslin RN, Saffran J, Newport E (1998) Computation of conditional probability statistics by 8-month-old infants. *Psychol Sci* 9: 321–324.

Aviezer H *et al.* (2012) Body cues, not facial expressions, discriminate between intense positive and negative emotions. *Science* 338: 1225–1229.

Block N, Fodor J (1972) What psychological states are not. *Philos Rev* 81: 159–181.

Bloom P (2002) *How Children Learn the Meanings of Words*. MIT Press, Cambridge.

Gosselin F, Spezio ML, Tranel D, Adolphs R (2011). Asymmetrical use of eye information from faces following unilateral amygdala damage. *Soc Cogn Affect Neurosci* 6(3): 330–337.

LeDoux J (1998) *The Emotional Brain*. Simon & Schuster, New York.

LeDoux J (2015) *Anxious: Using the Brain to Understand and Treat Fear and Anxiety*. Simon & Schuster, New York.

Nelson R (2006) *Biology of Aggression*. Oxford University Press, Oxford.

Pfaff D (1999) *Drive*. MIT Press, Cambridge, MA.

Pfaff D (2010) *Man and Woman: An Inside Story*. Oxford University Press, New York.

Saffran JR, Aslin RN, Newport EL (2013) Statistical learning by 8-month-old infants. *Science* 274: 1926–1928.

Schaafsma S, Pfaff D, Spunks R, Adolphs R (2015) Deconstructing and reconstructing theory of mind. *Trends Cogn Sci* 19: 65–79.

Searle JR (1980) Minds, brains and programs. *Behav Brain Sci* 3: 417–457.

Sommerville JA (2012) The development of fairness expectations and prosocial behavior in the second year of life. *Infancy* 18: 40–66.

Turing AM (1950) Computing machinery and intelligence. *Mind* 49: 433–460.

CHAPTER 6

Human–Robot Interactions

Robots will be used in very particular ways: a robot designed to supervise an elderly patient, and to ensure that she is fed and amused, will not conduct video surveillance in a war zone. The fundamental fact about robots is that they are task-specific; they are not all-purpose entities like human beings (though they can be reprogrammed and their functions altered/extended/diminished according to who may at the time be in charge). Robots do not therefore need to display the range of emotions that humans do. When mechanisms for emotion are programmed into robots (initially or at some later stage) engineers need only deploy those that correspond to the precise feelings that they want the robot to display. But while the range of mechanisms will be limited, each mechanism will not only need to be attuned to the robot's function, but each emotion must also be able to be expressed so that humans can interpret it. We will then have to learn how to respond in a way that makes sense in the situation.

The question is: What does it mean to "respond" to a robot, since robots will not be seeking our sympathy ("Oh, I am so sorry that your circuits are burning!"), and much less will they expect us to explain ourselves ("Oh, I thought that your circuits could tolerate the overload")? Robots, unlike people, will not expect that an emotional display on their part will elicit an emotional reaction from us. My argument in this chapter, as it is throughout, is that because robots perform specific tasks on which humans will depend, the human response must be calibrated to enable the robot to complete its task without harming itself, harming us, or a third person (or in fact some associated property). Our response will be purely functional, and will probably not feel like any emotion that we recognize.

It will be intended to make the robot perform or not perform some action, depending on what the robot's emotion communicates to us. Thus, just as robots will have a limited range of emotions, we will have a limited — and very likely scripted — array of appropriate responses. Our "relationship" with robots can therefore be learned, not just through experience (as we would with humans) but through a manual.

Humans interacting with robots need to know how robots "feel" about the task they are performing. If a robot is about to explode because one of its circuits is overloaded, it should be able to convey that it is extremely uncomfortable and that it is almost ready to collapse. The feelings expressed will need to be specific — by analogy, not just a generalized "pain," but a pain on the left side indicating a potential stroke. In the next chapter, addressing the legal consequences of robot–human interactions, we will discuss the problems caused by robots' nonspecific attempts to communicate their self-understanding to humans. But in our discussion below, we want to examine some types of robot–human interaction that are already possible or are soon to be, and then to speculate on the types of emotion that robots could likely display in these situations. As we have already noted, we think that based on our current understanding of human emotions, robots can be programmed to display the robot-equivalent (or, rather, -approximation or -counterpart) of what humans experience. Our capacity to succeed in this should become even more apparent as new types of robots — many of which are operated from a distance — increasingly perform tasks independent of continuous human control.

We should caution, however, that when we write about robots' displaying emotion, we are concerned with a spectrum of machines, some of which are more "sociable" than others. We are clearly concerned, for example, with the robot that feeds an invalid his or her breakfast, and may be programmed to communicate pleasure or dismay, depending on how the food is received. But we are also concerned with robots that interact with humans on a more remote basis, and that assist humans *subject to our careful direction* by performing tasks that machines can do more consistently and precisely. These might include medical and industrial robots, where the need for communicating a type of emotion will arise where something has gone wrong with the robot itself, and it can no longer perform according to its programmed objectives. What concerns us across this spectrum is that robots *in a situation* where they need to communicate are, in fact, able to do so. That need, of course, is measured by human benefit: if the human working alongside the robot is in danger from the robot (or is just prevented from completing their joint task), the robot should be able to express an emotion tantamount to fear, apprehension, or immediate concern. This is what any colleague would do. Likewise, if the robot itself is in danger, this impacts humans, and the robot should be able to communicate its self-concern.

When we speak of emotion in robots, we are not talking only of robots that look like humans. The vast majority do not. Rather, we are talking about those whose function puts them in contact with humans — steadily or intermittently, remotely or in the same space — and who will, for an array of reasons, need to interact with humans (to some degree) as humans would interact with each other. We are arguing that, in these situations, robots will need to express themselves in a way that we can understand as emotional. The emphasis is on understanding, a type of cross-species communication most readily accessible through emotional display. We know when dogs, cats, or bears are angry, even though they are not of our species, cannot speak, and usually pursue their own activities without much interest in us. The same can — and should — be true regarding robots, with whom we will be engaged in a lot more intricate, consequential tasks, even if only when the occasion presents. Robots of the near future will not just be servants or substitutes for humans, but actual partners, working alongside us in operations where each has a specialized function. In some cases, robots will do most of the work, leaving for us the inescapably human. The robots that concern us here already have a considerable degree of autonomy, and we can expect that as their expertise increases their autonomy will increase as well.

As an example that we can all understand, think of the cockpit of a modern jetliner. The information on display represents in *very large* measure the operations of a machine that virtually flies itself, minimizing human error. When the pilot does have to step in, however, he or she must quickly be able to interpret what the displays are seeking to convey. In these tense situations, accidents can occur and, as William Lange Wiesche points out, "many of them are now caused by confusion in the interface between the pilot and a semi-robotic machine." We are talking about this type of "interface," the communication modality between robots and humans that — at least in airplanes — can determine whether people live or die, or at least succeed in what they are attempting. What should this communication look like? In an airplane, as in any autonomous or semi-autonomous machine, increasing complexity designed to alleviate risk and error will only raise that question a notch further. As the late Earl Wiener once observed in one of his famous Wiener's laws, "Invention is the mother of necessity."

Whatever the solution, we must be enabled to trust it. As psychologists are beginning to understand, where machines are sufficiently complex we cease to trust what they are telling us in an emergency and fall back on our own instincts. That is to say, we are not sure that we can interpret them correctly when we ourselves are panicked or when the situation is somehow out of the ordinary. This can lead to the type of human error, founded perversely on either arrogance or insecurity, which we will need to alleviate.

One way of overcoming fear of this advanced technology is training to accept its superhuman capabilities. Here the armed forces have taken the lead. As *The New York Times* reported in May 2015, an "Autocollision Ground Avoidance System" saved a plane and its pilot in a combat mission against the Islamic State. The report noted that while the system warned the pilot, it did not take over his ultimate decision. He made that on his own, notwithstanding pilots' typical macho self-confidence.

What Robots Can Do

Robots are good at absorbing data, recognizing objects, and responding to information and objects. These broad categories encompass an array of tasks requiring intelligence; the trick, of course, is in how refined any particular robot's intelligence may be. That is to say, what sort of "information" can a robot recognize? Numbers are information, but so is the timbre of a human voice. Can a robot recognize that certain sounds constitute a human voice, a particular voice, or even a command? In some cases, this is already possible, so that if an identifiable individual told a robot to open a door, it would. Robots can recognize light and movement, and can be programmed to stop when there is a shadow indicating that a person has entered the sphere of their operation. On the other hand, it is doubtful that robots can yet direct traffic at a busy intersection. What is interesting, however, is how much robots can do in situations that involve interaction with humans.

Robots to the Rescue

Among the most recent developments is the role of robots in mine rescues. During the 20th Century, literally thousands of people died trying to save buried miners, often from asphyxiation or the collapse of fragile supports. Some died while trying to save children or explorers who descended into abandoned mines and could not return. Now, however, robots can go in ahead of rescue teams, report on conditions, and test whether the mine is safe for human entry. They can enter a mine as an advance team, probing for dangerous explosives, providing trapped miners with food and light, and removing dangerous material. They can provide detailed video images, and find out if trapped miners are still alive. (Many human rescue teams have risked their lives to perform a rescue, only to discover that the effort was in vain.) They can ferry supplies down and, in some cases, bring miners back up. In the mining industry, robots are now the first responders, working with human counterparts who direct them from a safe distance. The issue is: How safe is it for human rescuers to rely on robots' information, and for miners to rely on robots' attempts to help them? Put another way, what sort of emotions should we program into robots so that they can express concern for their own operability in conditions that may be flooded, overheated, corrosive, or impassable?

Well, what would *you* feel if you went down into a hot, flooded mine with the task of sending back pictures, as well as data on whether miners who had been trapped for hours still showed signs of life? You might, for example, be afraid and, if you had trouble breathing, you might panic. Most likely, you would not want the people above to rely on your radio transmissions, since the diminished oxygen supply was beginning to cloud your judgment; maybe smoke from an earlier fire was making visibility uncertain. If you were a robot, you would not need oxygen, but your video cameras might still cloud from the drifting smoke, and your mobility might be severely restricted by mud and debris. So, just as a human in this situation might fear for his or her life, and necessarily feel that his or her value as a reporter had been compromised, a robot might experience something similar: the equivalent of fear, or at least self-doubt, expressed as a concern for the continued utility of its video camera and, perhaps, for the rollers on which it could swivel 360 degrees.

An engineer, understanding how mine rescuers operate, would program a mine rescue robot to interact with its handlers outside the mine by sending signals that conveyed "fear" — and not just fear but, as with humans, an explicit type of fear that indicated what part of its mechanism was in danger or, in fact, was failing. Then the humans would understand that, at least for the time being, the robots should be recalled. They might further understand that robots that were ready to bring food down to the miners would have to be called off.

If the humans did not correctly interpret (or, in their desperation, just ignored) the robots' signals, and commanded the robots to continue moving forward, they might risk the robots' complete failure, such that they would be unable to turn around and ascend to the surface. Of course, a smart engineer would have given the robots a set of "secondary" emotions, to be called upon where their initial emotional display failed to communicate their condition. As in the case of a human mine rescuer, the robot might express a type of anger, a heightened emotional state that conveyed a principled resistance. It might express *stress*. It would be as if the robot were saying: "Look, I told you that the situation is no-go, and you are endangering me by insisting that I keep moving." It might send back electric pulses indicating dismay at its condition — not just physical, but with regard to its relationship with a human; in its operation manual, this would be explained as a profound warning not to exacerbate either. To do so would endanger the warranty and, at this point, the humans would be expected to get the message. If they still did not, perhaps the robot would be programmed to shut down altogether and wait for its own rescue by more sympathetic — or, rather, aware — operators. The initial operators might later be disciplined for endangering valuable robots and, ultimately, risking the lives of miners who might have been detected by these robots in a few more hours once the smoke had cleared and the mud run off.

Of course, if mine rescue seems like a very dramatic use for robots, it is not the only such use. Robots are employed in an array of first responder situations, notably in disaster relief, where they can deliver supplies, send back measurements of heat and wind, and assist humans in putting out fires or creating barriers. The issue here, as in mine rescue, is how effective a robot can be under extreme duress. If robots can go where humans cannot or should not, they may still have limitations that they will need to insist on when they are requested to push beyond them. Or even if they are not up against the limits of their performance, they may for any number of reasons need to convey that their performance is or about to be subpar. They will need to be able to transmit these feelings — these emotions of fear and self-doubt — so that they do not endanger humans who may be relying on them and even working alongside them. Moreover, where a robot is expressing fears for its performance, it would be putting many lives at risk to rely on it to nonetheless carry out a tricky rescue or even just a food drop.

As robots become less "out of the box" and more specifically responsive to individual specifications, it will be possible to build in the type of communication systems (and corresponding emotional signals) that will be most compatible with the situations they will find themselves in. We will be able to anticipate very particular types of crises, and create a type of sociability best suited to those crises, where the chance for missed signals is reduced to a minimum.

Robots and Doctors

While the average person may never be involved with a robot first responder, he or she may very likely experience robots that assist doctors in performing complex medical procedures. Such procedures might include prostatectomy, hysterectomy, cardiothoracic surgery, and ENT (ear, nose, and throat) operations. The robot works with the doctor to see and send back images at a minute level, and it allows doctors to use more precise, less invasive methods. Robots make specialized care more immediately accessible, since they permit off-site medical personnel to literally see, communicate with, and operate on a patient. Robots can also compound and dispense medicines, and help in rehabilitation. The medical robot is frequently an intermediary between doctor and patient, acting *with* the doctor *on* the patient. One can imagine an array of emotion-like expressions in such robots, ranging from encouragement to grave warning. But all such expressions, to be effective, will need to be highly specific. The robot will, ironically, be called upon to accurately diagnose itself.

For example, when it is assisting in a delicate operation and, for some reason, it detects that it cannot perform a particular movement with 100% reliability, it will need to communicate a very specific type of fear akin to "If I make this cut, I could miss the intended organ and injure the one adjacent." If its signal is missed, the doctor could be liable for malpractice; but if the signal is ambiguous, the patient could be hurt and there would no doubt be a fight over liability. The really tough question is: What do you do when you are halfway into an operation, the robot signals a vague sort of incapacity, and you do not how to correct for it? Doctors may feel that they need to be robot-doctors as well as doctors for humans, and the robot's on-site technician maybe on-call just as much as the emergency room staff. The point is that interacting with robots in hospitals, doctors' offices, and EMT vehicles may lead to situations that are just as stressful as any encountered by first responders in mining or other disasters. In fact, one can imagine that scenes of disaster will *also* be scenes where medical robots play a role, so that the stress of one naturally intensifies that of the other.

In fact, as robots proliferate, multiple types of robots will operate in the same space, along with their multiple handlers (nearby or remote). Stress will flow in many directions, with robots and humans attempting to carry out their respective tasks even as they potentially impede (or at least fail to understand) the others'. To the extent that one robot or human becomes entangled with another that is outside its or his line of reciprocal responsibility, there could be some very difficult attempts at communication. In such situations, there may need to be an overall director capable of reading multiple robots' emotional signals, and of directing the "traffic" among robots and humans so that the scene remains orderly and no one is hurt.

It is apparent that as robots begin to populate our world, any discussion of their operation will become geometrically more complicated, quickly becoming a sort of Robots 2.0 with regard to complicating factors that will need to be integrated into the ways that we engage them. It will not be enough to understand one type of robot or another — for example, the type of robot that we work with or that cares for our aging parent. Instead, and more fundamentally, we will need to understand how robots express themselves irrespective of the specific tasks that they have been programmed to perform. Thus, someone who works with mine rescue robots had better be able to figure out what a medical robot is attempting to convey, since in a disaster both types of robots could be on the ground. Just as our emotional expressions are fairly well standardized — fear is fear, whether in a child or an adult, a black or an Asian — so too engineers will need to develop signals that can be read by a variety of trained individuals *even if* they do not normally spend most of their time around the particular robots they will need to read.

As yet a further complication in this scenario, imagine the case where robots that are normally used in one capacity are called up in an emergency to serve in another. This could happen in an infectious disease outbreak where, for example, robots that feed and provide medicine to the elderly or bedridden are used to replace nurses and orderlies whose lives could be at risk. Or perhaps some emergency preparedness robots could be called in. Indeed, in October 2014, under the headline "Scientists Consider Repurposing Robots," *The New York Times* asked "Might robotic technologies deployed in rescue and disaster situations be quickly repurposed to help contain the Ebola epidemic?" In any such "repurposing," hospital workers may not be accustomed to working with *this type* of robot, and will need to pick up on its signals very quickly. When such robots indicate distress for any reason, they will need to have protocols for withdrawing the robots without causing contamination of everyone around them. The robots may be able to signal how they want to be withdrawn — perhaps after a self-sacrifice of some exposed limbs — and we will want to be able to efficiently comply.

Robots in the Workplace

While robots in a hospital or even at a disaster scene inhabit a doctor's "workplace," robots have made vast inroads into what is more conventionally regarded as a workplace, particularly the shop floor, where things are manufactured. Robotic arms help build cars and airplanes, and robots do the dangerous, repetitive work of making steel and tires. The fact is that industrial robots (a special category regulated by OSHA and subject to various industry standards) are what we usually think of when we think of robots, since a whole army of futurists is busy claiming that robots will replace manual labor (and, in many instances, the white collar workers who coordinate and supervise their work). But at this stage of robots' development, robots and humans very much work together in the shop. As we will discuss in the next chapter, this proximity can cause problems when a robot and its human coworker fail to perform as each has been taught to do (to program a robot on the shop floor is to "teach" it, according to the oddly anthropomorphic language that has come into use). Indeed, according to OSHA, there have been at least 33 robot-related deaths and injuries in workplaces ranging from bakeries and plastics factories to meat-packing plants, metal smelters, and auto-assembly lines.

Many of those incidents were gruesome, as where — in a car factory — a robot caught an employee on the back of the neck, and killed her when it pinned her head between itself and the part that she was welding. In the meat-packing facility, an employee accidentally activated a robot by stepping on a conveyor belt where robots were operating, and was killed when he became trapped. In an aluminum factory, a 150-pound ladle full of the molten metal pinned an employee against a wall and killed him when he sought to restart the pouring that had unexpectedly stopped. An employee in a plastics factory, who was trouble-shooting a robotic arm, died two weeks after the arm struck him in the head and ribs. Yet, if these sound awful, experts fear that the situation will become worse as the next generation of industrial robots comes on line. These will highly mobile, and no longer confined to cages or cordoned-off workspaces. They will become more like coworkers, doing work that takes brains, not just brawn, in spaces that we might never have expected to encounter a potentially dangerous nonhuman.

Because these robots are so widely in use, and are made by so many different companies — and because standards applying to one class may not apply to another, even though they serve similar purposes — it may be difficult to generalize about how well these robots communicate their self-diagnoses (or even their immediate local perceptions). Yet, if these robots are said to display artificial intelligence, then we can still ask: Do they have emotional intelligence, such that they can reflect on themselves sufficiently to understand what they should communicate? Moreover, do they, in a real sense, have the capacity to act as part of a community with the workers with whom they share space? When humans are part of a community, they look out for their neighbors; as colleagues in a workplace, they try to ensure that the environment is safe. Robots could be programmed to act communally, and to express fear *on behalf of* a coworker, be that a human or a robot. In situations where a human has crossed into a robot's field of operation, the robot will often just stop; but what about more general feelings of empathy, such that robots will positively seek to help a human (even if only by calling for help) who has gotten into trouble anywhere within its sight lines?

In my previous book, *The Altruistic Brain*, I wrote about specific "wiring" in the human brain that predisposes us to act generously, even selflessly, on behalf of other humans that we do not even know. Why then, since we can now describe this wiring, could a version of it not be introduced into robots, so that instead of merely evading harm robots might act to offer protection? Obviously, we cannot reproduce the complex neurohormonal attributes of humans, but we could still design robots so that when they detect danger on the shop floor they do not merely shut down (which could be sufficient in some instances) but also act, pushing someone out of the way or offering other rudimentary assistance (which could be necessary in others). A proactive industrial robot that communicates its concern and then *does something about it* should be the objective, so that the person who is involved understands what is going on and also is helped to survive the immediate danger. It might be that in communicating its concern, the robot actually signals the person to duck or flee — whatever it takes on the part of the robot to act in the spontaneously generous way that people normally would.

If a type of "artificial empathy" (corresponding to artificial intelligence) seems far-fetched, futuristic, or too anthropomorphic, I disagree. There will be a need for such emotional acuity in robots, and we know enough about the brain to make such empathy feasible. In the context of robots, it would not imply the sort of fuzzy notions epitomized in "I feel your pain." Rather, it would be a highly practical capability, enabling robots to work alongside people — and alongside people's property, that is, other robots — much as people would. The empathetic robot would not necessarily have contact with other people or robots, but would only be in a position of offering assistance, perhaps as minimal as a warning. Perhaps it could turn on the alternative power supply where the main power has suddenly gone down. Perhaps it could activate the fire extinguishers where it detects a fire, and then open the fire doors. The possibilities are endless, but they could be programmed to be situation-specific.

Where a robot was programmed with this type of empathy, the legal system is likely to kick in: What was the robot's responsibility, and did it fulfill it? If not, who should get sued? Would the robot's failure in this regard be insurable? At what cost? These are vexing questions that we touch on in the next chapter, although in the current stage of robots' development we think they are important to raise.

Robots in the Cockpit

By the time this book is in press, the Defense Department will have begun flight-testing its Aircrew Labor In-Cockpit Automation System (with the irresistible acronym ALIAS). The system is actually a robot that can be *seated* in the copilot's chair of a military aircraft to *act* as copilot. According to *The New York Times*, it will be able to speak, listen, manipulate flight controls that are built for humans, and read (and presumably react to) instrumentation. It will be able to land and take off, and could even take over from the pilot in an emergency. Equipped with voice-recognition and speech-synthesis capabilities, it could communicate with the pilot and with ground control.

As one Department official remarked, "This is really about how we can foster a new kind of automation structured around augmenting the human." Yes, at *this* stage. But at what stage does augmentation segue into a type of shared authority and, finally, outright replacement? Could a robot exercise better judgment than a human pilot, especially in those rare but crucial instances where no elaborate sensory apparatus could outdo the experience of a veteran flyer? These are questions that we will inevitably confront — and debate — as military technology begins filtering down into commercial aircraft. Will people (untrained by the military to deal with robotics) be willing to accept a nonhuman pilot in a flight, say, from New York to Beijing? Moreover, to the extent that robots do become active participants in any sort of flights — military or commercial — how will they be equipped to display emotion-like responses when, in a tight emergency situation, every second could be crucial in order to avoid a disaster?

Given the pressure on airlines to avoid human error (not to mention human frailty, such as that of the copilot who deliberately crashed his plane into the French Alps), robots flying airplanes could become immensely attractive. Their actual role(s) will vary, but their ability to communicate not just in technical terms but also at an emotion-like level will be crucial. A robot will need to be able to communicate fear, or perhaps even anger at the human pilot for an ostensibly "dumb" maneuver. It will have to be able to show discomfort, perhaps because it is overheated, so that the pilot and ground control will be alerted to take over its functions. Such capabilities can be devised, but they will have to be conscientiously installed so that human–robot interaction is effective and so that humans can be trained to rely on and accept their robotic counterparts.

Robots in Police Work

In the military, robots can disable bombs and perform some of the grunt work that used to fall to enlisted men. But most people will never encounter these robots. They may, however, be confronted by robots that assist the police. Such robots can now enter crime scenes and shoot at menacing individuals, shielding the police from direct and possibly dangerous contact with criminals. They can perform searches and uncover drugs. They can shine bright lights on a target and send back videos, allowing police to see inside secluded areas. In the next chapter, we will discuss litigation that has already arisen in which the use of robots at a crime scene was alleged to have abridged someone's constitutional rights.

The concern in any situation where a robot *replaces* a human police officer is how the robot, without all of the officer's training and human sensitivity, will change the dynamics of the situation just by being a nonhuman. Will the targeted person feel intimidated, and hence overreact? Will the robot, in other words, cause the situation to escalate not through any specific action but just by showing up? Will the police, because they may feel shielded and invulnerable, order the robot to take actions that a cautious policeman on the scene would avoid? Underlying all of these questions is the need for robots to be able to signal their intentions effectively, and for them to be able to read — and convey to the police — the intentions of the individuals that they encounter.

A large part of police work involves getting the suspect to trust the police, or at least not to resist their orders. This requires, in many instances, establishment of a certain rapport, even if only by means of eye contact. Will robots be competent to gain a suspect's confidence? The whole notion of "community policing," with the same cop on the same beat day after day, rests on this notion. Robots, on the other hand, may seem like the ultimate intrusion, an alien operative with no concern for anyone and with deadly accurate weapons.

Frequently, the police will use robots as loudspeakers, and merely amplify their own voices through a robot. But failure to actually be on the scene (even though they will receive video images of it) could, as we have suggested, radically distort the situation. The robot will be communicating with the police and with the suspect more or less simultaneously, and whether all of this information can be processed adequately at a distance is highly uncertain. What will a robot be able to convey about gestures, facial expressions, and expletives in a tense situation? (These are not always picked up on videos.) In crowd control, the uncertainties are multiplied. If a law enforcement robot is equipped to feel a type of anger, how will it express its feelings? How will its feelings be interpreted by someone already under duress, and willing to believe the worst just because of the robot's presence? Will the person immediately see such anger as a threat, and attempt to disable the robot (in the case that we discuss in Chapter 7, the robot *was* shot at and disabled).

In programming emotions into robots used in law enforcement, there will need to be a heightened awareness of how robots will alter a crime scene (or any potential disorder) because individuals may be afraid not only of the robots' power but of their limitations. Any emotional expressiveness programmed into robots will need to be both clear and immediately accessible to humans. It will be a challenge.

Domesticated Robots

"-bot" has become a short-hand add-on for an array of web-powered programs giving us such newfangled words as "spybot," "chatbot," "cleverbot," "lightbot," "botnet," and "Twitter-bot," to name but a few. While spybots, as their name implies, have little overtly to do with us, a chatbot and a cleverbot are highly interactive, answering questions and, in some instances, allowing corporations to keep tabs on their customers' thoughts. Yet, while all of these bots can intersect with an average person's life, they are too disembodied to concern us here. In the sense that they are not an apparent physical presence in our personal space, they are beyond the scope of this chapter. They are like the bots that trade stocks at lightning speed (and cause massive market gyrations), that crawl the web searching for data, and the botnets of communicating programs that (at their worst) send out spam or precipitate denial-of-service attacks. Just because we may personally use such a bot, and even become psychologically engaged with it, does not make it a "robot" rooted in our everyday reality.

On the other hand, everyday domesticated robots actually *inhabit* our space; they take up room; they can be said to have a type of localized brain or, more accurately, circuitry that distinguishes them (and confers a corporeality) different from web-based applications accessible from anywhere by multiple simultaneous users. In this sense, they are on the spectrum with people (albeit at the far end of it) in that they are individualized and physically apprehensible. We can imagine them as more than a brain (which is hard to do with a cleverbot since, even though it may be smart, we can hardly picture it as anything but a type of rarefied cyberspace intelligence).

One might argue about the status of dashboard-mounted GPS-based satellite navigation systems, which provide turn-by-turn instructions to drivers in dulcet tones. Yet, while people regularly swear by these things, they allow for no real interaction. We leave them to one side. We also put aside machines that reduce human functions to quantifiable outputs, but otherwise do not allow for real-time exchanges. For example, in September 2014, *The New York Times* ran a front-page story where the Thai government, incensed at bad foreign imitations of Thai food, described "an intelligent robot that measures smell and taste in food ingredients through sensor technology in order to measure taste like a food critic." After a prepared dish is inserted into the robot — perhaps a Thai curry, prepared by a foreign chef — the robot performs a number of chemical tests, compares the food to a predetermined quality standard, and rates the food numerically. Anything under 80 is inauthentic. Is this magic? A gimmick? In any case, it is a one-way street, and not of interest here.

Of course, with some tweaks it ultimately *could* be interactive, frowning at a dish, inviting the chef to add more ginger or Thai curry paste, and coaching him as he struggles toward a rating of 100. Maybe such machines will be programmable so that one's *own* preferences take precedence: once you have perfected a dish according to *your* taste, it will be the same every time irrespective of some "standard." No more standing over a hot stove, salting a soup five times until it is just right. Indeed, it is only a matter of time before robotic culinary coaches appear on kitchen countertops along with the Cuisinart machine (*Robo-Coupe* in French) and the remote-controlled coffee maker. You will be able to argue with the machine about the consistency of a buttercream or the lightness of a soufflé (it should be able to measure airiness), and discuss the best technique for one's own altitude (high-altitude baking is an art in itself). But this is yet to be.

Robots that do live with us include those that care for the young, the elderly, and the sick. They can perform simple tasks such as feeding and monitoring, and can be programmed to know our names. They can even read aloud from a thousand books programmed into their brains, and can cleverly converse much as a cleverbot can. The problem in this context is that such robots are in positions where people may develop an attachment, and may think that at some level it is reciprocated. That is to say, robots can foster the illusion that a relationship, like that among humans, is possible or has been achieved. If that illusion is disrupted, the person could be psychologically damaged. Indeed, the greater the dependency that develops — say, between a lonely shut-in and his or her caretaker robot — the riskier that relationship becomes. Thus, along with our development of domestic robots, we will need to develop a new branch of psychological counseling and troubleshooting, so that people remain aware that robots are only machines. This will be an especially tricky proposition, in that as we provide domestic robots with "emotions," there will be an even greater need to compensate at the back end for the illusions that may be created.

A domesticated robot that can express happiness when a sick person gets up and walks to the door — thus encouraging the person to keep on trying — is precisely the type of robot that also puts that person at risk for dependency. A robot that converses, and expresses delight when a person responds with a comeback, will have much the same effect. Robots that can protect old people and the sick from falling out of bed can produce psychological "side effects" that can exacerbate such people's fragility. As robots become more skilled, and their emotional expressiveness more refined, our strategies for dealing with them will need to evolve. For starters, we will need to instruct people who are served by such robots as to the nature of that service: it is not a human relationship. At the same time, we will need to ensure that in letting people know that the kindness and caring is an illusion, we are not abandoning them to machines. We will need to create a community around these people in which the robots are an apparent extension of *human* care, and not a replacement for it. In children and those with dementia, this may be difficult. But it will be necessary.

Of course, there are other types of domestic robots than those that provide care. Some are employed as dance partners, both for professional dancers (who need skilled, graceful partners that can respond to the tap of an arm) and for people in physical therapy. There is a robot called BabyX, an "interactive animated virtual infant prototype" which, as Jayson Greene observes, "learns exactly as a toddler does, mimicking facial expressions and responding to positive feedback." (Imagine her in parenting classes.) There are robot pets, which can offer the companionship of a puppy without requiring all of the care. The puppies can express a type of puppy exuberance and love, and can make children happy. Is this a problem? Probably not, except to the extent that these puppies suggest to their owners that companionship is a one-way street, and that we can be kept amused without having to give back in any way. Even goldfish, which are utterly indifferent to us, still need care, and we should not raise a generation of children who assume that delight comes at no cost to personal commitment.

The same principle applies to robots that will play games with us, pitching us tennis balls or softballs and giving us encouragement. To the extent that these robots stand in for coaches, we will need to ensure that people do not fall under any illusion of their concern for us. If we are not careful, our relationship with domestic robots *may become* like that with our smartphones — one of dependency — with the added complication that such robots are embodied and seem to be more than mere brains. Their emotions may not appear (at least to some) to be merely functional; they may instead seem personal and directed toward us as special individuals. Thus, just as we need to be on guard against industrial robots that can do us physical harm, we will need to be wary of domestic robots that may harm us psychologically.

If some sort of "personal" relationship with a robot seems fanciful — or, worse, just weirdly pathetic — think again. The toy industry has begun upping the ante on robots, much like the makers of medical devices, and has already introduced MiP, "Your New Robot Friend." This remarkable little character, powered by a battery and responding to guidance from an iPhone, an Android, or even a hand gesture, "is always trying to please." It has what the manufacturers call an "immersive personality," such that "Your new buddy likes to have fun, and invites you to join in. Show some praise, and MiP will be your new best friend, but push it down, and you're in for an angry robot!" Watching the slick video that advertises this little wonder, one is easily beguiled. Taking the tutorial is fascinating. But imagining MiP 2.0 (or even later incarnations) leads directly to all the concerns that we have just suggested. Robot playmates are not just video games, which no matter how lifelike are still two-dimensional and clearly just fictions. These machines could be just life-like enough so that we could become dependent. They already are able to display a type of anger, and we will need to learn how to register their emotions without becoming dangerously involved.

But personal robots are more than just pretend "friends." One of them could harm us physically, since they could feed us the wrong food; administer the wrong medicine; or hit us with a baseball going 60 miles an hour. They will therefore need to be equipped with the same capabilities for warning us — that is, for showing fear — when they sense that their functionality has been impaired. Most likely, just as there are current standards for industrial robots and OSHA regulations to ensure that such robots are employed correctly, we think there should be standards and regulations applicable to domestic robots. The Consumer Product Safety Commission (CPSC) has already issued recalls for certain robotic toys and lawnmowers, but we think this should be only the beginning. Much more sophisticated robots are on the scene now, and more are coming soon — robots that may or may not physically harm us, but that could easily have psychological side effects. The CPSC is probably not equipped on its own to deal with these robots, and may want to work with other federal agencies (for example, the National Institutes of Mental Health) or professional organizations like the American Psychological Association to map out a strategy.

"Professional" Robots

There has been a great deal of hand-wringing recently over robots' moving into the professions — law, accountancy, pharmacy, journalism, engineering, medical diagnostics — such that they will eliminate white-collar jobs. These are jobs that require not just substantial education but also sophisticated thought that traditionally has been considered uniquely human. The quintessential example is the robot that designs a robot that takes a human's job. Arguably, such hand-wringing is justified since, in the past, technological advances eliminated labor — blue-collar workers — but did not create wholesale redundancy of professionals. Now, however, this is a real possibility. Not only can robots produce journalism, scanning thousands of facts and weaving them into stories, but they can write symphonies and conduct them. They can create all manner of art, and use their "artificial intelligence" to be so deft, agile, and surprisingly original that they beat us at our own game.

It may be that the equation between man and machine must be run in reverse: rather than asking what robots *can* do — and extending the list toward infinite possibility — we will have to ask what they cannot. That latter category could, in time, become vanishingly small. It may be interesting to speculate as to whether there will be a type of Moore's law that could apply to robots, so that their progress toward human equivalence increases, and indeed accelerates with each generation. As robots become more like humans (which this book itself encourages), we will have to start thinking about what still distinguishes humans. Just because robots can *do* what we do does not, in every case, make them altogether like humans, since human skills are necessarily inflected by individual psychology (the result of accumulated experience having to do only partly with any particular skill). This is not to say, however, that the rush toward robot–human equivalence, however narrowly that equivalence is defined, will not result in a massive re-evaluation of what it means to be intrinsically human. While this enterprise may be a bonanza for philosophers — and possibly also law-makers — each individual will find himself or herself in situations demanding some personal accommodation with traditional notions of humanity. As the old Chinese curse says, "May you live in interesting times."

In the next chapter, we will explore some of the issues that we have raised here in a legal context. What is striking is that so many of these issues are only just beginning to be explored by the courts. We think that to the extent that at least some of these issues can be addressed outside of the courts — ideally, before the damage is done — this would be all to the good. Most of us have not yet begun living and working with complex machines that could hurt us and literally drive us crazy (even if our VCRs seem hopelessly user-unfriendly). But, as the human population ages, one of the most vulnerable segments of society may begin to do just that. More and more workers will encounter robots. Ironically, children already have. Not only should we equip robots with proto-emotional capabilities, but we will need to begin teaching everyone how to read robots and cohabit with them so as to minimize the risk.

Implicitly, robot researchers have begun to recognize that robots will need to resemble humans if they are successfully to cohabit with us. This is because humans have an affinity for their own shape, and will more likely collaborate with — and accept — robots that appear in some sense human-like. This includes being able to open doors and flip switches with recognizable hands. Of course, our argument dovetails with this perception, in that if robots can more easily transition into a home or workplace if they physically *appear* to "fit in," then that transition can be eased even further if robots express human-like emotions.

Yet there is always a catch. A recent *New York Times* article wondered whether making robots seem more human could possibly alienate humans from robots, since "these new faux-people... raise fundamental questions about what it means to be human" (October 29, 2013). If a robot dancing partner becomes a sort of Ginger Rogers (of whom it was said that she did everything Fred Astaire did, only backward in high heels), then all the more reason to enable that robot to express itself in ways that promote collaboration. Evolutionists understand that the human species succeeded because its members learned to cooperate (you could not hunt large animals on your own); we instinctively know how to do it. As the next generation of robots trundles off the assembly line, they should possess this capability, much of which depends on the capability to engage with others in ways that inspire reciprocity.

Further Reading

Goepfert RP, Liu C, Ryan WR (2015) Trans-oral robotic surgery and surgeon-performed trans-oral ultrasound for intraoperative location and excision of an isolated retropharyngeal lymph node metastasis of papillary thyroid carcinoma. *Am J Otolaryngol.* pii: S0196-0709(15)00097-6 (Epub ahead of print).

Van Dijk JD, van den Ende RP, Stramigioli S, Köchling M, Höss N (2015) Clinical pedicle screw accuracy and deviation from planning in robot-guided spine surgery. *Spine* (Epub ahead of print).

CHAPTER 7

Legal Implications of Living and Working with Robots

This chapter deals with the legal issues surrounding "sociable robots," that is, robots with whom we work and cohabit and with whom we have to communicate. (Notice the "whom," since robots are acquiring more human-like capabilities everyday.) All kinds of liability issues arise in this context, since who will be responsible when a robot performs imperfectly or causes harm? If in directing robots our communication fails (either because the robot failed to understand, or we spoke its language improperly), can we just apply the same principles of risk allocation that we do when dealing with people, or will new principles need to be created?

We think that at least some new thinking will be necessary, because even as robots assume the *roles* of humans, they still process information much more narrowly. Robots lack both empathy and socialization, and act out of pure logic — at which they will be very good, but which could be entirely inappropriate in a given situation (for example, if I order my colleague to perform an unaccustomed task he will not just eliminate me, but a robot might see that as the most logical solution to a problem). So, how can we build in communication mechanisms that will prevent this type of "cross-species" foul-up... and how will the law step in if they fail?

As a further complication, each person will vary in his or her abilities to interpret a robot's communication. How do we design a robot so that its signals are unmistakable? What if they are not? Should we only allow people to come into contact with robots after they have been trained to understand them? While this "training" approach may

163

work with standardized, industrial robots (although even these are becoming more and more customized), we do not think it will work in general, since robots will increasingly become a fixture of people's domestic situations, and most people rarely read, comprehend, or meticulously follow instructions that accompany their household machines. (How often have *you* followed the maintenance schedule on your smoke detector, and only changed the battery when the thing began beeping wildly?) As will be explained below, some important legal cases have arisen precisely because people failed to follow clear directions for how to deal with robots, even after these people received proper training.

Moreover, the real training deficit will emerge as a secondary market for robots, and robot parts, develops on eBay or similar sites. We see this development as inevitable. So, the issue is: How will this invitation to DIY types affect risk and liability? Should any such market be regulated, given the inherent risks? Or does the buyer simply assume the risk if he or she participates? It is not too far of a stretch to assume that the internet will empower everyone to become his own. Frankenstein, building robots from pieces of former robots — or just pieces of repurposed, maybe reconditioned machinery — without fully understanding how the pieces will work (or not work) together. Though there are industry standards for remanufacturing and rebuilding robots, they are unlikely to be followed by the average guy in a garage.

For example, suppose that a robot designed to transport packages on the shop floor is sold, maybe second-hand, to an old lady who wants it to carry her groceries. Her son adds a few gizmos from a robot fansite, and tells her it is ready to go. Suppose further that the old lady then instructs the robot to put down the packages while she rests on the way home. When a thief comes along and takes the packages, the robot may subsequently refuse to move because it was programmed to remain stationary except when making a delivery. When the old lady orders the robot to move without the packages, it resists; when she persists, however, shouting commands in an increasingly anxious voice, the robot assumes that it is being used for purposes that could create liability for the manufacturer and knocks the old lady down.

Who is at fault? The manufacturer, who built in self-protection that was too protective? The supplier of the robot's voice recognition software, who designed it to translate loud speech as a threat? The person who sold the robot second-hand? Or the old lady who bought it for an off-label use and then failed to understand how it worked? Was her son guilty of contributory negligence? After all, while he added features, he failed to disable one that caused the robot to do harm.

We have all heard of drug companies that promote drugs for off-label uses not approved by the FDA (for example, birth control hormones that coincidentally clear up your skin). If robots are sold for off-label use, does the manufacturer assume the risk? Do we? Suppose that some intermediary reprograms the robot and, like the old lady's son, adds a few more features. What if the manufacturer/seller provides instructions and we ignore them? Or perhaps if the robot warns us, and we persist, do we assume the risk? If the military sells off obsolete or "reconditioned" robots to make way for the next generation, can it sell them as is, with no warranty and with no responsibility for how they perform in the civilian world?

Should there be special legal requirements, comparable to legal doctrines of "strict liability," for robots dealing with the old, the sick, or the very young? Courts have begun to examine whether the sellers of robots — or those who control the robots' environment — should be held to a higher-than-normal standard of liability, and we will examine these cases. Yet a lot of these risks seem hard to quantify, or even to imagine. What happens, for example, if an old person, who becomes "attached" to a robot, discovers that the robot does not really care about her, even though it knows her name and makes eye contact? If this causes emotional distress, is this anyone's fault? (Probably not, under current standards.) Will mental health workers need to learn how to deal with the resulting disappointment? Can we even talk about offenses in this context, or are we entering some sort of brave new world where all of our standard, legalistic responses seem like square pegs straining in round holes?

Whatever we say about the possibilities of communication with robots, it will never match the reality of two humans talking (with eyes, hands, and vocal tonalities), and the gray area in between could be fraught. As robots become more like people, they also become — paradoxically — more difficult to categorize. It may seem like we can treat them as people, that is, as if they had emotions, but we will never be sure that robots' understanding of humans is commensurate with our understanding of them. Bridging the gap will be a continuing process, both in our day-to-day conversation with robots and at the level of research; even as we teach robots more about language, we will need to be sure that communication is effective in real time. We must never make the mistake that when a robot and a human appear to agree, they have reached their agreement through similar mental processes and can draw on the same data as they proceed with any undertaking. Robotic "consciousness" is likely to become a notion for philosophers to debate, even as engineers define it in precise terms involving signals and responsiveness.

With regard to the potential merger of robot and human sensoria — and sensibilities — there is already speculation about how we should treat robots that have been fitted (or, more ominously, "integrated") with human body parts. At what point do these cyborgs become eligible for civil rights reserved for human beings? Would we be required to give them a reasonable workweek (including time off), even as they displace human workers? Could they possibly be criminally liable, as some scholars have suggested? Questions like these seem more kaleidoscopic, more complex than any strictly legal analysis.

In this context, what happens to concepts like malpractice when robots take over law and medicine (robots are already drafting documents, performing legal "discovery," and making medical diagnoses)? The same question arises in engineering, social work, and architecture. Can you construct standards for robots' minimum performance in fields that require a combination of training, experience, and judgment — that is, the sort of "competence" requirements that lawyers and similar professionals must adhere to? Will robots need to be licensed to perform in their respective fields, and will they need to keep up with the latest developments in those fields? (If so, will we reprogram them, or will they be able to learn on their own?) Will there be the same sophisticated protocols for interacting with people?

Lawyers are required to be "ethical," to avoid conflicts, to put the client's interest ahead of their own and not bring opprobrium down on the profession. Can a similar regime apply to robots?

Because of the wide range of responsibilities that robots will assume, contractual questions, and those relating to potential liability, will inevitably arise. This chapter will survey some key developments in "robot law" (a field that does not yet officially exist) and will examine how it may need to develop once robots become more sophisticated — as well as more commonplace. Implicit in our discussion (though outside our immediate purview) are a host of vexing, long-range issues. For example, what terms will likely be written into union contracts to protect workers who could be made redundant by cheaper, faster robots? What responsibilities will workers have to upgrade their skills so that they can work alongside and be compatible with robots? Will some workers be required to come to a colleague's aid if he or she is menaced by a robot? Should they be trained and certified to do so, and be required to be on hand at all times? And how will any such "affirmative obligation" to provide assistance intersect with tort law, which holds that except where there are certain special relationships, no one is required to rescue another where the potential rescuer did not contribute to the risk? Is training the dividing line, or would an untrained coworker be liable to help just because he has the "relationship" of being a coworker? Some theorists hold that only individuals with the right training should be responsible for providing assistance to one who is in danger. Yet, even if there is such a proviso in a worker's contract, how will theories of responsibility be affected when the menace is a robot, and its responses are unpredictable? How far should anyone go (let alone have to go) to save someone else from a robot? Will that person be compensated if he is himself hurt? Will he be liable for damages if, in providing help, he botches the job?

Robots will make a lot of work for lawyers, and the field is already exploding. Just as patent lawyers are often trained engineers, robotics lawyers will require similar backgrounds as they toggle between liability law and the increasingly complicated demands placed on intricately designed (and often customized) robots. Adjacent fields, such as forensic experts capable of advising the court in complex legal cases, are expanding as well. Consider a recent Ohio case, *State ex rel. Camaco v. Robert J. Albu and the Industrial Commission of Ohio* (Ohio Ct. of Appeals, 2014), which featured an expert's opinion on whether a robot had a design flaw capable of causing an injury.

But let us step back a bit. From a broader perspective, will the fact that robots can work 24/7 influence the nature of "shifts," and will vacation or leave times be influenced by machines to which the concept does not apply? When robots are employed to do the same job that people do, how will that affect coverage of various laws (such as the heath law that covers companies with 50 or more people)? And what if one or many robots could be hacked? There are laws against hacking, but no law specifically contemplates robots as the target(s).

Though there is a host of issues that we cannot cover, we *will* pay attention to an area of extreme importance, namely medicine, where robots now fill prescriptions and help perform operations. These robots are already the subject of litigation, and they may become a subspecialty of lawyers now dealing with medical practice. Thousands of complaints against one robotic system have been filed with the FDA, and law firms advertise on their websites that they can help consumers win big settlements.

But perhaps the most troubling issue, which no court has yet faced, is where the introduction of robots causes us to face unaccustomed trade-offs in our very concept of fault and liability. Suppose that a kitchen robot is given ten pounds of apples to chop for a caterer's applesauce. A human cook would notice right away if any are rotten, and would just toss them away. But the robot chops them all and, as a result, the applesauce tastes musty and the caterer cannot use it. The robot did what it was supposed to do: chop apples. Under the best of circumstances (where all the apples were fine), the caterer would have been able to assign the cook to more demanding tasks, or might even have saved his salary altogether. Now, however, the caterer is out the cost of the apples, and maybe even late in catering for a children's birthday party. Can the caterer sue? After all, she got what she paid for in the robot, but the trade-off — that is, her saving on a cook's salary — may not have been worth the cost. Such failed trade-offs may lead to a legal vacuum, where people make choices and no one (or no thing) is to blame.

Yet, even in this context, where "trade-offs" are another name for market consequences, there is always the insurance industry to mitigate such consequences or, from another perspective, make them still more complicated. Suppose that the caterer had insurance against her robot's foul-up, and the cost of the discarded applesauce was reimbursed. Has the robot then proven its value? Moreover, if insurance covering the robot's performance is lower than the workman's compensation that the caterer pays, is the insurance an even better deal? After all, the cook might have cut his hand while chopping the apples, or slipped on the floor while hauling those apples out of the cellar. If insuring robots is *higher* than insuring people, then we have another set of trade-offs, each particular to the kind of business involved.

In the case of robots, as in the case of most new, disruptive technologies, so-called "legal" questions raise fundamental issues concerning values, and where we want society to be headed so that people benefit from such technologies (I resist the terms "coexist" and "live with," which imply a type of concession to the inevitable that I do not accept). In fact, American courts have been thinking about robots in the workplace, and even outside it, for a generation, and both industry and government have issued rules giving rise to a new class of experts: those whose job is to "teach" robots (yes, that's the term) and to monitor their performance in industrial environments. What have the courts said about robots, and how should that make us think about emerging legal issues that will only become more complicated as the technology develops?

Some context. On a few occasions, the Supreme Court has alluded to robots, but in no case with such far-reaching implications as in *A Book Named "John Cleland's Memoirs of a Woman of Pleasure" v. Attorney General of Massachusetts* (1966), involving the state's ban on a sexually explicit 18th Century novel commonly known as *Fanny Hill*. In that case, Justice Brennan sought to liberate Americans from standardized, unthinking behaviors that he associated with mechanization, that is, with robots: "Decadence, in a nation or an individual, arises not because there is a lack of ability to distinguish between morality and immorality, but because the opportunity for self-expression has been so controlled or strangled that the society or the person becomes a robot." Brennan juxtaposed sexually independent individuals (like those glorified in *Fanny Hill*) with those who endure the preordained and conventionally "moral" lives endorsed by Norman Vincent Peale's *Sin, Sex, and Self-Control* (the title tells it all). The latter example, he concluded, does not give scope to one's humanity — does not allow one to "express" oneself — and so gives rise to a robotic state.

On the basis of this logic, the Court reversed Massachusetts' ban on the novel, helping in its own way to kick off the Sexual Revolution. *Blade Runner* (1982), where robots have sex with people and truly challenge humans' monopoly on sexual freedom, was still a generation away. From the Supreme Court's point of view in the 1960s, robots lack autonomy and cannot express themselves, especially in relationship with another (entity? person?).

Of course, sooner or later cases rising to the Supreme Court from lower courts will challenge this position, so that the doctrine attributed by the Court to *Fanny Hill* will (in some yet indeterminate measure) apply to robots as well. In fact, I am going to use the Court's position on robots as an ironic segue to the rest of this chapter, indicating that while "robots" may still be synonymous with entities that cannot be expressive, in reality it is just the reverse. Even as cases (many of them!) use "robot" in the manner underscored by the Supreme Court, as where a murderer is said to act like a robot and show no feeling, other cases actually deal with robots interacting with people and necessarily communicating with them. So there is a paradox where, in colloquial and general terms, we think of robots as zombies, while in actuality robots are getting more like people. Sooner or later, casual references to robots, as well as their use as foils in ringing legal endorsements of freedom, are going to have to change.

Liability for causing work-related harm. *Payne v. ABB Flexible Automation* (8th Circuit, 1997) was the first reported case where a robot caused the death of a human being. Of course, "causation" in a legal sense, with all its financial consequences, differs from what we mean by saying "That robot pinned a human being down, and as a result he died." The mere fact that a robot is involved, and even does something awful, does not in itself establish liability. In *Payne*, the court had to determine whether the robot's manufacturer "proximately" caused the death — in this instance, by omitting a presence-sensing safety feature — or whether Payne's negligence could have been the cause and that other causes could not be ruled out either. In ruling in favor of the manufacturer, the court allowed the first contested death-by-robot to go uncompensated.

The case is interesting because it represents one court's early attempt to apply existing legal principles to robot–human interaction. Here is how the court describes what happened: "On the evening of September 27, 1994, Payne was working as a 'cell operator' at [his place of employment]. As a cell operator, Payne was responsible for operating and programming an M93 IRB 6000 automated robot used for the production of aluminum automobile wheels, and for supervising other employees working in the robot's cell. There were no witnesses to the accident. Payne had instructed his coworkers to take a break, while he remained in the cell. When a coworker returned to the cell, he found Payne pinned between the robot's gripper arm and a wheel inside a drilling machine. Payne died two days later." What went wrong, and whose fault was it?

Payne's wife alleged that ABB was negligent and strictly liable for designing and manufacturing a robot that was defective and unreasonably dangerous. Indeed, the manufacturer admitted that it had not installed a "presence-sensing device" which would have enabled the robot to detect entry of personnel into its sensing field. Nonetheless, the court stated: "Hindsight knowledge that the presence-sensing device might have possibly prevented the accident, in and of itself... does not establish that the robot was defective." It went on to explain that not all defects create liability: "A 'defective condition' is a 'condition of a product that renders it unsafe for reasonably foreseeable use and consumption.' A product is deemed to be unreasonably dangerous when it creates a danger which is *beyond that which would be contemplated by the ordinary and reasonable user of the product who possesses the ordinary knowledge of similar users regarding the risks, hazards and proper uses of the product.*" (Emphasis supplied.) In other words, Payne — who possessed the "ordinary knowledge of similar users" — should have known how to properly use this robot, and should have appreciated the risks of *not* properly using it.

Moreover, not only had Payne's wife failed to prove that ABB was negligent, but she had also failed to *disprove* that there might have been other causes of death. Until she could eliminate them, she simply had no case. In fact, there *was* evidence that Payne himself had been negligent, since he entered the robot's cell before "locking it out," and without following his employer's guidelines regarding the robot's speed.

So, in many ways, the court applied a traditional products liability analysis to a very nontraditional working relationship: that of a robot and a human. Replace the robot with a chainsaw (some models also now come with sensors), and you could get the same result. Should we be satisfied, therefore? Or will there need to be industry standards (and, indeed, OSHA requirements) for types of warning systems and communication capabilities? Should there be fail-safe modalities that depend on the robot's ability to signal danger, to "worry" about its operator and ensure that he or she understands and reacts to the impending danger? (Mr. Payne was deemed by the court to have understood the danger, but no one was making sure that he reacted to it.) In other words, should we give robots the simulacra of workplace emotions — the same as Mr. Payne's absent colleagues might have displayed — so that workers can interact with robots more or less as they would with anyone who puts them in harm's way? A robot that expresses something in the nature of fear, and insists that its apparent concern be addressed, is a robot that could save lives. Though OSHA does not consider audible/visible warning systems as a substitute for "positive" systems that limit access to a robot or shut it down in an emergency, it does recommend redundant safety measures, and here is where expressive robots could come into their own.

If such capabilities were regularly installed in robots, the practice could influence some courts. For example, the court in *Payne* noted that "while not conclusive, evidence that a particular safety device is commonly used in a particular industry carries weight in determining whether a proper standard of care has been breached." But Payne's wife failed to introduce any such evidence (probably because there was not much in the 1990s). There *was* evidence that the robot failed to meet the American National Standard for Industrial Robots and Robot Systems-Safety Requirements 15.06-1992 with regard to slow speed definition. But the court held that this was irrelevant, since the plaintiff did not produce evidence that the robot was operating at slow speed at the time of the accident. The American National Standards Institute (ANSI) publishes important standards for robot manufacture and performance, and it would be good to see the Institute focus on the problem of robot–human communication. ANSI standards are referenced by OSHA regulations, but no one has yet to augment this regime with basic neuroscience.

Recent cases demonstrate that industry is becoming serious about safety, and that it will not tolerate lapses in procedures designed to protect workers from robots. For example, *Shawn Orr v. Orbis Corp. and Menasha Corp.* (N.D. Georgia, 2010) concerned the dismissal of a setup team leader tasked with ensuring that his employer's robots operated safely and effectively. The court upheld his dismissal because the employee failed to follow the company's strict safety procedures, endangering not just himself but his coworkers. From the company's point of view, it was obviously worried about its liability where, knowing that a worker had not followed procedures, it might have been found to have condoned the lapse. Also, it might have been worried about its insurance, since policies frequently require that shop floor safety procedures be followed and documented. As robots become more pervasive, will there be new clauses in such policies that impose special duties and heightened care on the insured? We can envision training and safety protocols that involve establishing workers' proficiency in human–robot communication, and that require supervisory workers to instruct their team members in how — as a group — they will respond to robotic signals involving group safety.

Yet, as the cases demonstrate, no matter what safety precautions are in place, employees may seek to circumvent them. Consider

Miller v. Rubbermaid (Ct. of Appeals, Ohio, 2007), where nobody understood why an employee who was "teaching" a robot evaded every safety protocol, crawling under a fence and in between plastic injection molds to teach the robot while it was (1) in continuous operation mode and (2) experiencing a problem with a program designed to prevent the employee from dealing with the robot while it was in that mode. When a suit was brought alleging that the company caused the employee's death, there was testimony that the employee should have waited for the maintenance technician, and not done anything until he fixed the problem. Not surprisingly, the court ruled in favor of the company, holding that where an employee operates so far beyond the bounds of safety rules his employer cannot be held liable. Of course, the spooky part of this case is precisely the question that was left unanswered: Why was this employee so impetuous?

I would like to speculate. Could it be that people feel under pressure to produce — and perhaps to meet quotas/expectations — now that they have enhanced tools such as robots? Even though, in *Miller*, it was company policy not to place oneself at risk, there may have been a certain internalized pressure to do exactly that. Even where we are not competing *with* robots, we are competing *against* a standard that machines allow employers to set for us (think of Charlie Chaplin's *Modern Times*). It is almost as if, in *Miller*, the plaintiff was daring a robot to "get" him when the plaintiff refused to play by the rules. This is a type of insult, plausible perhaps in the instant because of the extant pressure (even if the robot will not take your job, someone else might). But, whatever the pressure, we need to treat robots as if they *could* be insulted, and then avoid doing it — perhaps at the robots' own prompting, if not at our own. Robots need to be built with systems that can register insults, and then warn people not to continue. When every other safety protocol fails, there may still be an opportunity for one-on-one exchange, not unlike that which a much stronger person feels when boys start calling names.

Yet, even apart from the workplace, what is interesting is that not just employers, but anyone that operates premises where robots are in use, may be responsible for maintaining an environment that is safe for robot–human interaction. As a consequence, they may demand that robots and humans be able to communicate directly. Consider *Rodriguez v. Brooks Pari-Automation et al.* (N.D. Texas, 2003), where a worker invited on the premises to install a robot suffered serious injury. Here is how the court described what happened: "Plaintiff alleges that he and Towns [a coworker] were not furnished with communication devices while installing the robot in the elevator shaft and, as result, they had to yell up and down the three-story elevator shaft to communicate with each other. Plaintiff crawled into the elevator shaft to yell instructions at Towns. As he crawled out of the shaft, Towns engaged the robot, and it ascended at a high rate of speed towards Plaintiff, who was unable to stop the robot, because there was no emergency stop inside the shaft. To avoid the robot, Plaintiff attempted to hang onto the side of the elevator shaft with his left hand, and as he was holding on, the gears of the robot ran over his left thumb, severing it at the lower knuckle... As result of the injury, he suffered nerve damage, numbness, pain and impairment in his left hand." The court sent the case back for reconsideration since, it said, Rodriguez could likely show that the premises owner knew or should have known about the unsafe condition, and that it did nothing to alleviate the condition. But suppose this robot had been equipped with the same capacity to "worry" that we discussed above in connection with *Payne*, as well as a capacity to communicate directly with Rodriguez? Would not this have made a difference? If robots are going to populate more and more of our environment, then this type of interactivity needs to become increasingly common.

Medical device liability. Robotic medical devices are now preferred to their manual counterparts in an array of complex, delicate procedures, notably prostatectomy (removal of the prostate gland to prevent the spread of cancer). A manufacturer's liability for these devices' malfunction is based on the same products liability principles that we have already discussed, but because such malfunction arises in a hospital setting — where the operators are highly trained doctors — there is less likelihood of human error. There is thus greater pressure on the manufacturer to show that it was not at fault when someone claims injury. Yet, in perhaps the best-known case in this area, *Mracek v. Bryn Mawr* Hospital and Intuitive Surgical Inc. (E.D. Pa., 2009, aff'd, 3rd Cir., 2010), the manufacturer easily prevailed. The court demanded a high standard of proof from the plaintiff as to the inherent defectiveness of the robot, observing that a surgical robot is more technical and complex than even a car bumper, which the Third Circuit Court of Appeals has held requires expert testimony to establish a design defect. In other words, a surgical robot is not your average machine, which most jurors can understand, but rather a high-tech piece of equipment that only initiates can assess. In *Mracek*, the plaintiff adduced no expert testimony that the robot was inherently defective, and so he lost. As the court explained, the machine could have worked well on any number of occasions; the fact that it failed in Mracek's case does not prove anything.

Mracek also failed on a theory of breach of warranty, since he failed to eliminate all the other possible causes of his injury. To the average person, this might seem unreasonable, since after his prostatectomy Mracek experienced erectile dysfunction and groin pain, which he had not experienced before. But, as the manufacturer asserted, "this is a matter of complex surgical innovation — a robotic procedure requiring a multitude of medical decisions and judgments made by the surgeon. The use and timing of various ancillary medical equipment in connection with this innovative and complex procedure reinforces that any number of reasonable secondary causes could [have been] or were responsible for the alleged damages." In light of this rationale (which obviously impressed the court) it is likely that the use of surgical robots will always create problems for aggrieved patients, since both medical judgment and the vagaries of the human body can never be second-guessed with precision.

Mracek based his case on the fact that at several times during his prostatectomy the robot flashed error messages when a company technician was summoned, even he could not diagnose the problem, and ultimately the robot was abandoned in favor of a manual instrument. But, even with the devastating outcome, the argument failed to satisfy the legal standard for liability. So the issue is — at least for our purposes — whether the error messages could have conveyed more specifically why this particular robot was "upset," that is, what was it *actually* feeling in terms of its self-diagnosis that caused it to display those messages? Could there have been more communication with the robot? Should robots be designed to be more forthcoming about their "feelings," and should we be better educated to read them? If there is any takeaway from the *Mracek* case, it is that even robots that are well-built and are properly maintained can sometimes fail, especially under pressure, and so we need to cooperate more effectively with the robots themselves. We need to make it possible for robots to cooperate by treating them as if they had feelings, that is, concerns about their own operability. It is somewhat ironic that while Mr. Mracek was able to tell his doctors the nature of his ailment — both through tests and by talking to them — no one expected the same level of revelation from the robot. Perhaps it is time that we did.

Indeed, the law is moving sufficiently fast that there is likely to be no choice. Currently, there are dozens of cases outstanding against the manufacturer of a surgical robot. As these cases wend their way toward settlement, economics will itself enforce major changes in how robots and humans interact.

Law enforcement. As robots and violent offenders confront each other, law enforcement just is not what it used to be. Imagine reading a complex court decision putting away someone for 52 years on four counts of shooting at a peace officer, and encountering this: "To restart communication with appellant [who was armed and dangerous, and had gone silent inside a house], Assistant Sherriff Tom Gattie deployed a robot at around 8:00 p.m. The robot stood approximately two-and-a-half feet tall and three feet across. It had tracks on the bottom for mobility, four adjustable lights and several cameras mounted on top for visibility, and a microphone and speaker for communication. It was operated by a toggle switch. It held in its claw a window punch for breaking glass. A 12-gauge automatic shotgun was mounted on the robot." All of a sudden, your expectations of a tough but conventional legal slog are shattered, and you wonder: Can this be real? Yes. The case, extraordinary in its revelation of how robots are used by the police, but also asserted in legal arguments against them, is *Reinhardt v. Feller* (E. D. Cal., 2008). It is a window into the future.

Among the many issues in the case was whether Reinhardt could claim self-defense when he shot at the police after the robot was sent in. He argued that use of the robot was such an "escalation" by the police, that there was no way he could have withdrawn. That is to say, he had to defend himself. From the court's description of this menacing piece of machinery, the argument might seem plausible. But it failed. There was no evidence that the robot actually fired at Reinhardt and, indeed, Reinhardt disabled the robot by shooting at it once it reached a door and turned on its lights. The court noted that there really was no escalation, since the police were *already* aiming at the house — the robot (quickly disabled) was redundant — and that "sending in the robot was not so sudden or unanticipated as to prevent appellant from withdrawing safely." It took the robot 30 to 45 minutes to penetrate the house, and during that interval "appellant could have informed the police that he was withdrawing or surrendering." So, faced with a robot with a gun and a window punch, there was still reason enough to withdraw.

Of course, what is interesting to us here is the possible effect of robots on the law of self-defense. The court is saying that Reinhardt should not have been scared, and should have put down his arms. But suppose that he *was* scared, just because the thing *is* a robot and we might all be scared of a semi-automatic miniature tank. Therefore, how should designers of robots, working with law enforcement and informed by the courts, deal with the inherently intimidating appearance of robots? As robots enter into law enforcement, we need to think about such issues. Should a robot display "empathy" for a person that it confronts, and give him a warning: "Look, I am giving you five minutes to put your gun down, because I am not a menace and you have lost your right to claim self-defense." That is to say, should the robot realize that it heightens the apparent threat level just by being nonhuman, and therefore compensate by displaying "emotions" that are readable by a human?

Police tactics may need to take into account the human–robot interaction, even though — as in *Reinhardt* — the robot was a virtual microphone for the police, with no independent speech of its own. Merely confronting a robot sets up layers of concern that confronting an armed human being does not. Is a policeman's speech *coming* from a robot automatically amplified, not just in decibels but in threat level? To the extent that robots acquire the ability to issue commands on their own, will these need to be couched in such a way that the person receiving those commands does not immediately freak out, perhaps becoming so paralyzed that he cannot follow those commands even if he wants to? Will there need to be some general knowledge in the community — perhaps imparted through community education projects — on how to deal with robots in tense situations?

What sort of offense is it to shoot at robots: damage to property? Suppose that an injured robot is programmed to shoot back, that is, to defend itself. How would *that* affect the doctrine of self-defense? If a robot kills, is that judged according to the same standards as if the police kill? Suppose that someone throws down their gun and says "I surrender." Can he be sure that the robot will believe him? A policeman in the same situation would be able to interpret the person's face, gestures, and tone of voice — but would a robot be able to? Suppose that a robot gets someone in a chokehold, and he screams that he cannot breathe. Would the robot understand? What would the robot do? A policeman (at least in New York) knows not to engage in chokeholds, and *should* know to loosen up if someone is struggling for breath. Is doing battle with a robot exactly like fighting with the police?

The issues are as extensive as the policeman's rule book, and are heightened even further when law enforcement takes place in a prison. Robots could easily be prison guards, but the need for proper communication with prisoners (who still have certain civil rights) would need to be programmed into them. Given the brutality of some prison guards, this could be a real advance. But, as with everything else having to do with robots and the law, there will be an array of difficulties before robots reduce the likelihood of litigation as opposed to increasing it.

Indeed, there are types of cases that have not yet even been brought. These could involve, for example, robots in our homes or in nursing homes. When the first driverless car crashes into a crowd, who will be liable? When a robot becomes your tennis coach, or your sparring partner in the gym, could you sue for breach of contract or ask for a reduction in your annual fee if your performance fails to improve? And what about robot first responders? Will the government be liable if one of them causes injury or destroys valuable property? (The Department of Homeland Security and the National Institute of Standards and Technology already sponsor elaborate testing standards for "response robots" that can be "rapidly deployed," "mobile in complex environments," and that are "equipped with operational safeguards.") Could a local government be bankrupted if its robotic first responder corps causes unanticipated harm? The possible legal disputes are infinite, since we do not even know how robots will inhabit — or control — our space in the coming years.

We do know, however, that it will be necessary to equip robots with a type of sensitivity to the environment that humans take for granted in themselves. Failing that, a lot could be lost. As a Canadian judge observed in *Canada Post Corp. v. Canadian Union of Postal Workers* (British Columbia, 2013), "On the day companies begin to operate solely with robots, they will lose the valuable contribution of a worker thinking about improvements to be made to the production system." The idea is to prevent that type of loss by enabling robots and humans to make each other maximally efficient.

CHAPTER 8

Robots Tomorrow

While rapid advances in robotics are often based on amazing science, there is no similar support as to predictions of future progress, which are frequently amazing but not always scientific. Below, I summarize some of those predictions — which often contradict each other — and offer my views as to which seem plausible from a neuroscience perspective.

Wading through all the contradictions, I still see nothing that would refute my main argument that we will in the future need to treat robots *as though* they had emotions. Following that, it seems inevitable that human interactions with robots will become routine and develop synergies.

To say this another way, I will argue for a view of robots that would further enable "the human use of human beings" even as robots encompass more and more of what humans used to do. This phrase, initially coined by MIT professor Norbert Wiener in *The Human Use of Human Beings: Cybernetics and Society* (1950, revised 1954), does not tell us exactly how to achieve maximum human potential in an automated world. It is a sentiment, not a formula. Nevertheless, Wiener had a compelling vision: to free humans from the bonds of age-old, repetitive drudgery by enabling them to *cooperate* with machines, giving humans the time and energy for more creative work. Without being presumptuous, I want to riff on that vision, taking from it the general idea that we can hand on accustomed tasks to robots even as we amplify our identity as humans. I want to suggest that robots will not just eliminate human jobs, shoving us further toward the margins of useful activity, but also — and as a consequence — enable us to experience our humanity in ways that most of us lack the tools or are just too busy to do.

In making this case, I realize that most neuroscientists would not dare try to define what it means to be human so as to apply Wiener's phrase. But I do not really think that is necessary. Rather, I will argue that it will be essential for humans themselves (not some exalted philosopher) to imagine new activities and jobs beyond what robots can do. I will argue that our relationship with robots can help us in that task. If we are liberated to become more creative, our first creative act must be to figure out how we construct our relationship with robots so that we feel vivified, not deadened by it. Such an approach may clear an economically supportable way for robots to perform an increasing share of other tasks necessary for human existence.

If, as I have argued in Chapters 4 and 5, robots will be programmable to act *as though* they had human emotions — a premise of this book — then where do robots end, so to speak, and humans begin? One part of the answer, as I have just suggested, lies in robot–human relationships. By analogy to the many types of purely human relationships — utilitarian, reproductive, and romantic relations — robot–human interactions will be overwhelmingly utilitarian. Knowing that we are involved with machines programmed to be useful will allow people to perceive the difference between the human and the robot mentality as the two work side by side. Humans will have a broad-gauge versatility that robots designed for specific purposes will not, however cleverly they are engineered. Humans will push their own limits. They will attempt to operate beyond their comfort levels and, in the process, will define their "use" as (creative, daring, individualized) humans going forward. One can easily argue for robots acting smart, as I do below. Robots acting social is a bit more controversial. Yet I choose to be optimistic, notwithstanding the array of scholarly perspectives on the future of social robots (some of which see them as agents of economic collapse for everyone except those at the very top). Robots will, indeed, take routine jobs but humans, being humans, will imagine higher levels of jobs that will make society better and richer.

Smart, Yes

Everyone knows that robots, equipped with "artificial intelligence" circuitry, can perform certain intellectual tasks at a high level. Consider chess-playing machines and IBM's Watson winning the TV game Jeopardy. But do we know anything more than that?

Yes. For one thing, the Defense Advanced Research Projects Agency (DARPA) has sponsored the creation of robots that can be trained to do a wide variety of tasks, from opening doors and using tools to climbing stairs and clambering over debris. Because of massive amounts of computational power due in part to artificial intelligence circuitries working in parallel, robot capacities are growing fast — for example, an MIT team facing the DARPA challenge included 12 students who were juggling DARPA preparations with final exams. Their robot, Helios, did well, but messed up a wood clearance path and fell off a ladder. The winner, from the University of Tokyo, did all assigned tasks smoothly but "with deliberation." Many artificial intelligence experts using the computational power available to them think that they would do best mimicking neural circuits in the human cerebral cortex.

The best brain-and-behavior scientists, like Andrew Meltzoff at the University of Washington and Terry Sejnowski at the Salk Institute, understand that learning can be understood as a computational problem. Learning by children and by robots (a) involves inferences from the structures of their environment and series of events in their environment, and (b) the calculation of transitional probabilities, whether from word to word or act to act. As these two have argued, statistical regularities and covariations in the world thus provide a richer source of information than previously thought. Even more exciting is social learning, where the simplest step is imitation. Up to now robots can perform all of these learning steps quite well as young children. In Meltzoff's and Sejnowski's words, "machine learning algorithms are being developed that allow robots and computers to learn autonomously."

On the other hand, informed scholars can argue back and forth between the wildest, most optimistic dreams for robot learning and the direst warnings. For example, Emily Anthes has stated: "As robots insinuate themselves ever more deeply into our lives, understanding their limitations will be as crucial as knowing their capabilities." In addition, "when robots do things we don't understand, like follow rules we don't know, we [should] wrest control away from them, even when they are right."

Most important, Sejnowski distinguishes between "traditional approaches to artificial intelligence, where the goal is to create a cognitive machine that creates a model of the world and computes responses based on that model" and normal brain physiology. With its limited capacity, the brain selects only the most important sensory inputs to process and the most effective responses to store. That is to say, even though artificial intelligence products and computational neuroscience inform and inspire each other, they are fundamentally different. Robotic futures will benefit from both. More than that, Murali Doraiswamy, Professor at Duke University, reminds us that as robots get smarter they will be even better at *enhancing* human capacities. As noted in Chapters 2 and 3, the differences between robot intelligence and human intelligence — referred to as well by T. Sejnowski — allow them to *synergize* all the more easily. Our peak abilities differ from robots' combined, we should do better than ever.

As we compare robotic intellectual capacities with the performance of the human brain, in scientific history we can go all the way back to Hermann von Helmholtz, the 19th Century scientist who fought against vitalism and instead used the empirical, reductionist approach neuroscientists know today. Vitalists sought what they called "the vital principle," a metaphysical idea. Helmholtz, by contrast, did the kind of physiological experiments on the visual system that we still can recognize as valuable today. He saw the basis of psychology in physiology. For the origins of artificial intelligence we go back to the 19th Century mathematician Charles Babbage. As James Gleick writes in *The Information*, Babbage's Difference Engine could perform a wide variety of calculations automatically.

Over the decades, machines improved, but Gleick takes special note of MIT's Vannevar Bush, who in 1945 came up with his

Differential Analyzer. Bush's accomplishment can obviously be compared with Turing's, described in Chapter 1. In Gleick's words, Bush's Differential Analyzer "did not manipulate numbers. It worked on quantities — generating curves. We would say now that it was analog rather than digital. *Its wheels and disks were arranged to produce a physical analog of differential equations*" (emphasis mine). While mathematicians/engineers such as Bush were producing machines, the electrical engineer Claude Shannon was producing the fundamental theory of artificial intelligence or, indeed, information of any sort. Having graduated from MIT, Shannon, working at Bell Labs, needed to measure the transmission of intelligence. In 1948, he published his theory which consisted of a simple equation with only four mathematical terms, an equation that amounted to a universal measurement of *im*probability, *un*expectedness, surprise. Thus, a whirlwind tour of some of the intellectual highpoints developed for reductionistic neuroscience, artificial intelligence and their intersections.

What does this have to do with robots? Technology writer Clive Thompson argues that in our future interactions with robots, the products of artificial intelligence — whether we are business leaders or neuroscientists — will make us smarter. As I pointed out in Chapters 2 and 3, the fact that peak human abilities will often complement peak robotic abilities suggests that the opportunities to synergize will be legion. Computers — vehicles of artificial intelligence — allow different creative people to communicate more easily with each other, thus allowing them to spark new ideas and search for multiplicative advances.

In John Markoff's view, "the limitations [of artificial intelligence] are simply those of an understanding of the objects to be attained, and of the potentialities of each stage of the processes by which they are to be attained, and of our power to make logically determined combinations of those processes to achieve our ends. Roughly speaking, if we can do anything in a clear and intelligible way, we can do it by machine." I agree with Markoff when he says "we can be humble and live a good life with the aid of the machines, or we can be arrogant and die."

Social Robot Behaviors

Hopes run high for the development of robots that can participate fully in a wide range of social interactions with humans. For them to do so will require high levels of what engineers and neural network designers call "artificial intelligence." Here I offer examples that conflate novel artificial intelligence with that of actual robots in order to examine (a) how far robot designers will get in developing social behaviors, and (b) their economic consequences.

As mentioned earlier, smaller, safer robots make better coworkers. One could hardly overemphasize the impact of future robots on the workplace. According to *The Wall Street Journal*, the number of industrial robot installations has gone up about fivefold during the past 20 years. Are they simply going to release humans to do better, higher-paying jobs, or are they, in the journal's words, "going to devour jobs that required the uniquely human skills of judgment and dexterity"? I examine both sides of this high-stakes debate.

On the one hand, as mentioned below, the geopolitician George Friedman is not worried. He thinks that human populations will be low enough that we will be grateful to have robots doing many jobs.

On the other hand, MIT professor Erik Brynjolfsson has stated: "It's gotten easier to substitute machines for many kinds of labor. It could happen that there are people who want to work but can't" (because robots took their and others' jobs). The computers are getting better and better. They are not just doing laundry, they are doing legal research. Going forward, the dynamics will be complicated and, in my view, unpredictable. DARPA's Siri, capable of "smart aleck humor," as stated in the *Technology Review*, is described by MIT's Boris Katz as having artificial intelligence "not all that sophisticated." University of California–San Diego's RUBI can recognize some emotional facial expressions, cry if its arm is pulled hard, and can teach Finnish to a young child. Jonathan Gratch at the Institute for Creative Technologies in Los Angeles has created a "robot-psychiatrist, Ellie," to whom patients could disclose personal information more freely than to a real doctor. The point is that on a number of campuses with a variety of intellectual aims, what passes for social intelligence in a robot is coming on strong.

Rather than trying to define a "social robot" as MIT professor Cynthia Brazeal does in *Designing Social Robots*, I would simply focus on easily recognized social behaviors for their own sake. Robots will be able to interact with us and even communicate with us in a socially efficient manner. Brazeal has already gotten part of the way there with her robot Kismet. She especially wanted to achieve human–robot interactions that would include infant/caregiver exchanges. Not only did she have to construct environments that would permit Kismet to behave as desired, but also Kismet's circuitry had to be, in Brazeal's words, "capable of regulating the complexity of its interactions with the world and its caregiver."

I was especially interested in how Brazeal built "motivation" into Kismet's system, because of my own expertise in hypothalamic neurobiology and all the motivational states that it controls. For this she built in homeostatic mechanisms by which a "need state" is recognized in the robot's circuitry and, in most cases, need is reduced by consummatory behaviors (acts that terminate a goal-directed behavior, such as eating). Especially important for the main argument of this book was that Brazeal was able to build in behaviors that resembled emotive behaviors. Easy to understand is a "fear-like" reaction, since all Kismet has to do is to "move away from a potentially dangerous stimulus." Brazeal understands that the nature of fear reactions depends on the fundamental level of arousal of the system. So her robot's "fear" responses range from fatigue to "anger."

Kismet's proto-social skills have been a work in progress, but Brazeal does not lack ambition along these lines. She hopes to have robots that can "correctly attribute beliefs, goals perceptions, feelings and desires" to others. In social neuroscience, such abilities sometimes go under the rubric "theory of mind." As mentioned, my postdoctoral fellow Sara Schaafsma and I recently criticized this field of work for having vague terminology and sloppy thinking, but it is clear that from an early age human infants can intuit others' perceptions and desires; Brazeal wants her robots to do so, too.

Looking ahead, Brazeal can fine-tune robotic abilities on which she already has concentrated: Kismet should give recognizable social cues; Kismet can "perceive and appropriately respond" to cues offered by humans; perhaps the robot and the human can actually adapt to each other and Kismet will use these abilities eventually as a basis for potential social learning. In addition, as I hinted, Brazeal's aims are not modest: she uses the idea of "robot personality," personal recognition, and empathy.

Robots' entry into the world of social behavior is being facilitated by two new developments in the physical and biological sciences. First, neuroscientists know more than ever before about the neural circuitry active during certain types of social behavior. As reviewed by Timothy Behrens at the Oxford Centre for Functional MRI of the Brain, at least two types of circuits are emerging and more are to be expected. Neurons in circuits that regulate social preferences are closely related to reward circuitry, including cells in the basal forebrain and the ventromedial prefrontal cortex. But neurons involved in understanding other people's social intentions are located differently, including the dorsomedial prefrontal cortex and the posterior part of the brain near the juncture of the temporal lobe and the parietal lobe. The more we know about how these circuits work, the easier it will be to imitate some of them in robotic social controls.

Second, interactions between social beings ("agents") permit, in the words of Icosystem's Eric Bonabeau, "agent-based modeling, which relies on the owner of computers to explore dynamics out of the reach of pure mathematical methods." Agent-based modeling can be used to predict social interactions when, in Bonabeau's terms, "the complexity of differential equations [describing those interactions] increases *exponentially* as the complexity of [individual] behavior increases" (emphasis mine). Nonlinear, discontinuous behaviors — the most difficult kinds to be handled — can be submitted to agent-based modeling techniques. The bottom line is that powerful computer modeling techniques not only make it more likely that we can envision the social behaviors of an individual robot, but also make it more likely that we can understand, ahead of time, the social behaviors of groups in which robots participate.

But here is a warning: MIT professor Sherry Turkle emphasizes the downside of celebrating increased man–machine interactions. Her plenary lecture at the 2013 annual meeting of the American Association for the Advancement of Sciences was titled "The Robotic Moment: What Do We Forget When We Talk to Machines." Her recent book was titled *Alone Together: Why We Expect More of Technology and Less From Each Other.* I am reminded of her perspective when I walk along Central Park South in New York City and see a couple at brunch — each attending solipsistically to his or her cell phone. Turkle worries that new technologies actually will get in the way of genuine human-to-human connections. She has two reasons to be disturbed, seemingly opposite to each other in how they work to fray human friendship. First, the person attends to the machine rather than to another person. Second, the creation of a false "hyperpersonal effect" — online connectivity that cannot actually be sustained in real life.

In Turkle's interview with Mark Fischetti for *Scientific American,* she says that in an era of social technologies "people have a tremendous lack of tolerance for being alone," and that "people start to view other people as objects." One implication is that at least some of us will suffer a loss of empathy. Although in our book *The Altruistic Brain* we argue that the human brain — even as linguist Noam Chomsky says that it is wired to produce grammatical sentences — is also wired to produce altruistic behavior, Turkle seems to think that robotic participation in society will erode empathic instincts. For example, if a robot "could stand in as a companion for an older person," that older person could not "tell the story of their life to something that has no idea what life is or what loss is." I have argued in earlier chapters that robotic abilities may synergize with human abilities precisely because they spring from different mechanisms, but for Turkle "this is the question before us: in losing human–human functioning for better chances to succeed, are we undermining or are we enhancing our competitive advantage" when we bring robotic social behaviors online?

In support of Turkle's point of view is a paradox: there are data describing (a) how children are spending so much time peering at screens that they are losing the ability to recognize emotional signals that emanate from people's faces, and (b) how machines are getting better than people at deciphering people's emotional cues. This is a stunning paradox. Scholars like Patricia Greenfield, Professor of Psychology at UCLA, and the late Clifford Nass (Stanford) worry that as we become more dependent on machines, we become less like people, even as we build machines that become more like people. The old categories are breaking down. My optimistic response, as a student of the brain and behavior, says that this development is an extension of human capacities and will foster productive synergistic relations between people and robots. In the long run, only a small proportion of the population will get confused as to the objects of their interactions.

Threats or Blessings?

Most of us are worried that increasing robotic competence in the workplace will cost humans their jobs. Thus, *The Economist* warned about "computers that can do your job and eat your lunch." Economists Henry Siu (at the University of British Columbia) and Nir Jaimovich (Duke University) said "Many of the routine occupations that were once commonplace have begun to disappear, while others have become obsolete. This is because the tasks involved in these occupations, by their nature, are prime candidates to be performed by new technologies." And machines are becoming capable of performing more sophisticated tasks, like legal research searches and language translation. Google has used machines for unsupervised learning.

Worse, some economists have used sophisticated probabilistic statistics based on the major components of demographic change — fertility and life expectancy — to project changes in in the world's population. In the estimate of a large group at the United Nations, "the world population is unlikely to stop growing this century. There is an 80% probability that world population, now 7.2 billion people, will increase to between 9.6 billion and 12.3 billion in 2100." Clearly, if the population on the planet increases and if job numbers decrease because of robots in the workplace, then a lot of people will not be able to get jobs. Bad news.

Others take the opposite view. George Friedman, an expert in geopolitics, holds a much more optimistic position in *The Next 100 Years*. For starters, consider that the world's water supplies are certain to diminish. Therefore, it is an advantage that robots do not drink or have to wash daily. About population and jobs, first, he thinks that transformations of roles within the family and promotion of women to ever-greater levels of responsibility will encourage what he calls a "population bust." Advanced industrial countries like Germany, France, Japan, and Russia will see population declines (a trend already begun). Other countries that have been growing fast will, in the second half of this century, according to Friedman, "see their populations stabilize." Even African countries like Congo and countries like Bangladesh will have a lower rate of population growth. Friedman may be indulging in "pie in the sky," but by his thinking populations will be small enough that jobs will be available.

Which of these extreme views is correct? Paraphrasing applied mathematician Joel Cohen, Professor at Rockefeller and author of *How Many People Can the Earth Support*, nobody knows for sure. How could we possibly know the fertility choices of our grandchildren's children? In Cohen's view and that of experts at the Population Council, we are not doing enough now to promote education, especially that of women. This major factor combined with others will determine the magnitude of the robot–human jobs problem in the coming decades.

Second, while robots may do routine jobs, I believe that we will be able to devise many better jobs that will await human job seekers. We noted in Chapter 6 that increasing numbers of elderly people need help. As UCSF professor Louise Aronson has pointed out, robots don't need sleep, don't need to be fed (and, as mentioned, don't need daily watering), and can get a lot of household tasks done while the elderly patient sleeps. Filling roles like this, robots don't assume levels of creativity unique to humans. As I argued in Chapters 2 and 3, the complementarity of constraints on the abilities of humans and robots renders them ideal for synergies as they go about their complementary tasks. At airports, for example, artificial systems of all sorts can handle routine tasks, but people are required to sort out myriad travel difficulties for harried passengers.

The Human Use of Human Beings

Long before most of us began to work, the computer visionary Norbert Wiener — "founder of the science of cybernetics" — thought about the relationship between computers and the human nervous system. As noted below, his phrase "the human use of human beings" required that computers be used for mechanical tasks so that humans would be freed to make value judgments. In fact, humans would be *required* to make value judgments — moral choices between good and evil.

A professor of mathematics at MIT, Wiener was able both to put artificial intelligence in historical context and to peer into the future. Regarding historical context, consider the abacus. As summarized by John Markoff, "in the abacus we carry out a human intervention of exactly the same sort as in combining numbers on paper, but in this case the numbers are represented by the positions of balls along a wire rather than by pen or pencil marks." A mechanical system here represents a human thought. And, sure, so does the desk computer. For Wiener, the relationship with machines was all about communication and control. As he would say, it is all about cybernetics.

Cybernetics. Think in terms of dynamic systems. The word "cybernetics" comes from the Greek word "*kubernetes*" ("steersman") or "*kybernao*" ("to steer"), from which we get our word "governor." Cybernetics comes into play when we analyze or exploit control and communication loops that offer useful feedback to the main initiating element, be that in a machine or a person. The basic idea can be boiled down to information: I communicate information to some other entity about action A and when that entity "communicates back to me he returns a related message which contains information primarily accessible to him and not to me." In that way, guided by feedback, action A will come closer to its goal than if I had acted alone. Looking back in history, Wiener couched his ideas in the work of mathematicians Fermat and Leibnitz, who were interested in automata. Looking forward, Wiener envisioned that "society can only be understood through a study of the message and the communication facilities which belong to it; and that in the future development of these messages and communication facilities, messages between man and machines, between machines and man, and between machine and machine are destined to play an ever-increasing part." That was in 1954!

For Wiener, "the physical identity of an individual does not consist in the matter of which it is made... the biological individuality of an organism seems to lie in a certain continuity of process, and in the memory by the organism of the effects of its past development. This appears to hold also of its mental development." He compared that biological individuality to computers — read artificial intelligence and robots. Thus, "in terms of the computing machine, the individuality of a mind lies in the retention of its earlier tapings and memories, and in its continued development along lines already laid out."

Ultimately, Wiener wanted to envision the developmental direction of computers and their impact on human society. As other thinkers have done, he recounted the history of the first industrial revolution and how that changed the nature of work. Machine use allowed society to move away from "the most brutal forms of human labor." What about future computers, artificial intelligence and robots? Wiener saw that "the machine plays no favorites between manual labor and white-collar labor. Thus, the possible fields into which the new industrial revolution is likely to penetrate are very extensive, and include all labor performing judgments of a low level."

As I read Wiener, he wanted us judiciously to consider the application of each new technology. In the ancient world: "The Greeks regarded the act of discovering fire with very split emotions. On the one hand, fire was for them, as for us a great benefit to all humanity. On the otherhand, the carrying down of fire from heaven to earth was a defiance of the gods." Do we shrink when we face scary new possibilities, in his case, the splitting of the atom, in our case, the applications of nuclear energy, for example? No, we will move forward thoughtfully "with fear and trembling. He will not leap in where angels fear to tread, unless he is prepared to accept the punishment of the fallen angels." This is Wiener's somewhat romantic wording.

Here is the key point: "Neither will he calmly transfer to the machine made in his own image the responsibility for his choice of good and evil." Machines, be they computers, artificial intelligence nets, or robots, will always be completely literal-minded. We will always need to know its laws of action based on our design. Then ethical choices between good and evil must be left up to us. Wiener did not say that we should be *limited* to thinking about good and evil. Nor is it the case that we all must become priests, rabbis, or professors of ethical philosophy. Rather, in the constantly shifting divisions of labor between robots and humans, Wiener and others would insist that there will always be higher mental and emotional functions that will be uniquely human. We will need to deploy those faculties, thus to keep creating our own possibilities. We will need to stay one step ahead of the machines but, in my view, the necessary impetus will be what human–robotic synergy is all about.

Wiener's dictum urging "the human use of human beings" requires that we leave all ethical decisions to ourselves. Even when we build fragments of ethical systems into robots, humans make the decisions about what to do and how to do it. This holds whether we conduct those decisions one by one in solitude or program those decisions into robots. It is an urgent matter. In Wiener's final words on the subject, "the hour is very late and the choice of good and evil knocks at our door." Again, a somewhat florid statement of a computer scientist's outlook which, stated more moderately, could be applied to thinking about robots.

Hence, we return to the work of Ronald Arkin discussed in Chapter 1. He could program robots to behave better than some human soldiers might (some of the time) because he could appeal to the clearly stated rules of war. His ambitious titles include words like "robot missions with performance guarantees" and "recognizing situations that demand trust." And his field of design engineering is expanding such that Professor John G. Taylor at King's College London, writing in the journal *Cognition and Computation*, could claim to have achieved cognitive control architecture for perceptions and actions that is based on actual human visual pathways and "allows it [the robot] to learn and reason about how to behave in response to complex goals in a complex world."

One Bottom Line

In parts of this book I intertwine ethical feelings and decisions with emotions. Such thinking fits the tradition of the "moral sentiments" of Immanual Kant and other classical thinkers. To face the general problem of robots behaving well in society, we need to have a system of ethics. Thus, in turn, emotions must play into the question of how one behaves well in society.

I argued in Chapter 5 that recent progress in the neurobiological understanding of emotional mechanisms will allow us to mimic the relevant forebrain circuitry in robots. That done, we will have to treat robots *as though* they had emotions. The next step is easy. I argued at length in my previous book *The Altruistic Brain: How We Get to Be Naturally Good* that neural wiring for producing altruistic behavior can be clearly understood. Thus, it can be mimicked. The mechanisms that permit us to produce altruistic behaviors include the anticipation of our next act, the perception of the person (or robot) toward whom we will act, and the merging of that percept with our image of ourselves. Downstream in that circuitry, decisions will be made. With all of these mechanisms built in, it is inevitable that we will be treating robots as we treat ourselves.

I am not talking here about the "inner nature" of robots — just their objectively defined behaviors. That is to say, I have tried to avoid sounding like a philosopher. Consider the verbiage of philosopher Thomas Nagel's, "What is it like to be a bat?": "Apart from its own interest, a phenomenology that is in this sense objective may permit questions about the physical basis of experience to assume a more intelligible form. Aspects of subjective experience that admitted this kind of objective description might be better candidates for objective explanations of a more familiar sort. But whether or not this guess is correct, it seems unlikely that any physical theory of mind can be contemplated until more thought has been given to the general problem of subjective and objective. Otherwise we cannot even pose the mind–body problem without sidestepping it."

Likewise, in *The Future of the Brain*, New York University's philosophy professors Ned Block and Gary Marcus disappointed me by getting tied up in their own verbiage. I have tried to avoid that. Scholars like the professor of philosophy at Oxford University, Nick Bostrom, worry that if the artificial intelligence embodied in advanced robots finds some way to enhance itself, then, in an uncontrolled fashion, they could do real damage to us. As far as I know, it is impossible to prove that could never, ever happen. His answer, and others', is to make sure that computers are programmed so that, indeed, they do what we want them to, and nothing else. Exactly how to accomplish such a plan will need a lot of thought by computer scientists and mathematicians. The sooner, the better.

Outlook

The old categories of humans and machines are breaking down. Humans are interacting more and more with devices that use artificial intelligence. Conversely, as I argued in Chapters 4 and 5, robots will become better and better at para-emotional interactions with humans. In Chapters 2 and 3, I wrote that, in terms of sensory and motor mechanisms, robotic developments will not closely mimic the human brain. So, look for chances for synergistic and especially productive relations between humans and robots.

As a practical scientist I do not think you need to be on the side of techno-utopianism in order to see the bright side of human–robot relations. Guruduth Banavar, chief of cognitive computing at IBM, would say that yes, there are risks in building more artificial intelligence into robots, but that there are also risks in *not* doing so. My optimistic outlook (stated above) says, yes, robots will take some human jobs. But humans, being humans, will innovate and redesign jobs for humans for the future. That is to say, do not look for robots to be visionary. Humans will be visionary, using cognitive faculties that depend on the human cerebral cortex and that neuroscientists currently are struggling to understand.

Further Reading

Behrens T *et al.* (2009) The computation of social behavior. *Science* **324**: 1160–1168.

Bonabeau E (2002) Agent-based modeling: methods and techniques for simulating human systems. *Proc Natl Acad Sci USA* **99**(Suppl. 3): 7280–7287.

Bostrom N (2014) *Superintelligence: Paths, Dangers, Strategies*. Oxford University Press, Oxford.

Breazeal C (2002) *Designing Social Robots*. MIT Press, Cambridge, MA.

Cohen J (1995) *How Many People Can the Earth Support?* Norton, New York.

Cutsuridis V, Taylor J (2013) A cognitive control architecture for the perception–action cycle in robots and agents. *Cogn Comput* **5**: 383–395.

Friedman G (2009) *The Next 100 Years*. Anchor, Random, New York.

Gerland P *et al.* (2014) World population stabilization unlikely this century. *Science* **346**: 234–239.

Gleick J (2011) *The Information*. Pantheon Books, New York.

Hof R (2013) Deep learning. *MIT Technol Rev* **116**: 30–36.

Meltzoff A, Kuhl P, Movellan J, Sejnowski T (2009) Foundations for a new science of learning. *Science* **325**: 284–288.

Meulders M (2010) *Helmholtz: From Enlightenment to Neuroscience*. MIT Press, Cambridge, MA.

Pfaff D (2006) *Brain Arousal and Information Theory*. Harvard University Press, Cambridge, MA.

Pfaff D (2014) *The Altruistic Brain: How We Are Naturally Good*. Oxford University Press, New York.

Schaafsma SM, Pfaff DW, Spunt RP, Adolphs R (2015) Deconstructing and reconstructing theory of mind. *Trends Cogn Sci* **19**: 65–72.

Sejnowski T (2010) Interview with P. Nair. *Proc Natl Acad Sci* **107**: 20601–20606.

Thompson C (2013) *Smarter Than You Think*. Penguin, New York.

Turkle S (2011) *Alone Together: Why We Expect More of Technology and Less from Each Other*. Basic Books, New York.

Wiener N (1954) *The Human Use of Human Beings*. DaCapo, Plenum, New York.

Index

Schiff, N., 79, 80
Searle, J., 128
self-defense, 178, 179
sex, 110–112, 117, 120, 121
sociable, 142
spinal cord, 69, 70, 72–75, 77, 78, 81, 82, 84, 90, 93, 94
stress, 146, 148

Tomasello, M., 136
Turing, A., 13, 19–21, 26
Turkle, S., 189

vision, 35–37, 39, 41, 43–45, 51, 55, 60, 64, 66
voice recognition, 165

Wiener, N., 181, 182, 192–194
workplace, 150, 151, 161, 169, 172, 174